Problem Organisms in Water: Identification and Treatment

AWWA MANUAL M7

Third Edition

American Water Works Association

Science and Technology

AWWA unites the drinking water community by developing and distributing authoritative scientific and technological knowledge. Through its members, AWWA develops industry standards for products and processes that advance public health and safety. AWWA also provides quality improvement programs for water and wastewater utilities.

MANUAL OF WATER SUPPLY PRACTICES—M7, Third Edition

Problem Organisms in Water: Identification and Treatment

Project Manager/Copy Editor: Melissa Christensen
Production Editor: Carol Stearns

Library of Congress Cataloging-in-Publication Data

Problem organisms in water : identification and treatment.--3rd ed.
p. cm. -- (AWWA manual ; M7)
"Replaces AWWA manual M7, Animals associated with potable water supplies: operators' identification guide (1966) by William M. Ingram and Alfred F. Bartsch"--Foreword.
Includes bibliographical references and index.
ISBN 1-58321-292-2
1. Aquatic pests--Identification. 2. Aquatic pests--Control. I. Ingram, William Marcus, 1913- Animals associated with potable water supplies. II. American Water Works Association. III. Series.

TD491 .A49 no. M7 2003
[TD427.A68]
628.1 s--dc22
[628.1'61]

2003060084

Printed in the United States of America

American Water Works Association
6666 West Quincy Avenue
Denver, CO 80235-3098

ISBN 1-58321-292-2

Printed on recycled paper

Contents

List of Figures, v

List of Tables, ix

Foreword, xi

Acknowledgments, xiii

Introduction, xv

Chapter 1 Actinomycetes 1

Biology and Ecology, 1
Significance for Water Supplies, 4
Control Strategies, 5
References, 5

Chapter 2 Iron Bacteria 7

Biology and Ecology, 7
Significance for Water Supplies, 11
Control Strategies, 12
References, 15

Chapter 3 Sulfur Bacteria 17

Biology and Ecology, 17
Significance for Water Supplies, 20
Control Strategies, 21
References, 22

Chapter 4 Nitrifying Bacteria 23

Biology and Ecology, 23
Significance for Water Supplies, 25
Control Strategies, 26
References, 27

Chapter 5 Nematodes 29

Biology and Ecology, 29
Significance for Water Supplies, 32
Control Strategies, 32
References, 33

Chapter 6 Bloodworms or Midges (Chironomid Larvae) 35

Biology and Ecology, 35
Significance for Water Supplies, 37
Control Strategies, 38
References, 39

Chapter 7 Crustacea 41
Biology and Ecology, 41
Significance for Water Supplies, 45
Control Strategies, 45
References, 46

Chapter 8 Rotifers 47
Biology and Ecology, 47
Significance for Water Supplies, 49
Control Strategies, 49
References, 50

Chapter 9 Zebra Mussels 51
Biology and Ecology, 51
Significance for Water Supplies, 53
Control Strategies, 54
References, 56

Chapter 10 Algae 57
Biology and Ecology, 57
Significance for Water Supplies, 57
Algal Divisions, 59
Control Strategies, 66
References, 66

Chapter 11 Protozoa 69
Biology and Ecology, 70
Significance for Water Supplies, 76
Control Strategies, 77
References, 77

Appendix A AWWA Survey on Nuisance Organisms, 79

Appendix B Troubleshooting Guide for Problem Organisms, 87

Appendix C Suggestions for Optimizing Conventional Water Treatment, 91

Appendix D Actinomycete Culture Agars, 95

Appendix E Iron Bacteria Presence/Absence and Quantity Methods, 99

Appendix F Color Section, 101

Abbreviations, 137

Index, 139

List of AWWA Manuals, 145

Figures

1-1 Actinomycetes, phase contrast 1,000×, 2

1-2 Actinomycetes, dark-field illumination, 400×, 3

1-3 Bacterial colonies—typical colony versus actinomycete colony, 50×, 3

2-1 Single-celled iron bacterium *Siderocapsa treubii*, 8

2-2 Filaments of *Sphaerotilus natans*, 9

2-3 Laboratory culture of *Gallionella ferruginea*, 9

2-4 Stalk fragments of *Gallionella ferruginea* and iron–manganese precipitate, 10

2-5 Filamentous iron bacteria recovered from well water. DIC microscopy, 1,000×, 10

2-6 Partly filled impeller channel, 11

2-7 Largely clogged bowl volutes, 12

2-8 Column pipe with iron biofilm, 12

2-9 Clean versus clogged impellers, 13

3-1 Side view of a tubercle, 18

3-2 Cross section of a slime, 18

3-3 Cross section of a tubercle, 18

3-4 An SRB *Desulfovibrio desulfuricans*, 20

3-5 *Beggiatoa,* a colorless sulfur bacterium, 20

3-6 An operating water well being pumped, influence (suppressive) on SRB activity (theoretical), 21

3-7 Operational water well during period of quiescence (not pumping), causing the SRB activity to increase and the hydrogen sulfide to spread (theoretical), 22

4-1 Electron micrograph showing cross section of a nitrifying bacterium, 24

4-2 Water quality parameters during a 1985 nitrification episode at Garvey Reservoir, Los Angeles County, Calif., 26

5-1 Roundworms (phylum Nematoda), 30

5-2 Nematode in bright-field, 200×, 31

5-3 Nematode at 250×, 32

6-1 Two-winged flies (order Diptera), 36

6-2 Pupa of a chironomid (Insecta, Chironomidae), 20×, dorsal view, 37

6-3 Parthenogenetically formed eggs in abdomen of a chironomid pupa (Insecta, Chironomidae), ventral view, 70×, 39

7-1 Crustaceans, 42

7-2 Crustaceans, 43

7-3 Crustacean *Bosmina* with eggs, iodine stain, bright-field microscopy, 60×, 44

7-4 *Asellus aquaticus* (Crustacea, Isopoda), dorsal view, 5×, 45

7-5 The amphipod *Hyallela azteca* (1.0 cm) isolated from a point-of-use filter, 46

8-1 Rotifers (phylum Rotifera), 48

8-2 Rotifer, 200×, 49

9-1 Life cycle of the zebra mussel, 52

9-2 Adult zebra mussel, 53

9-3 Zebra mussels encrusted on structure, 53

9-4 Veliger larvae of the zebra mussel, showing the velum extended at lower right, 54

9-5 Impact of zebra mussels on screens, 54

10-1 Algae, Chrysophyte, *Dinobryon* (3 cells), and others, unstained bright-field, 240×, 62

10-2 Algae, Diatom, *Asterionella*, unstained bright-field microscopy, 120×, 63

11-1 *Arcella* (0.1 mm), 70

11-2 *Paramecium* (0.2 mm), 70

11-3 *Giardia* cysts and *Cryptosporidium* oocysts stained by immunofluorescent antibody, 4',6-diamidino-2-phenylindole (DAPI), and Nomarski differential interference contrast (DIC) microscopy, 72

F-1 Iron bacteria. *Leptothrix* (note sheath), 1,000×, 102

F-2 Iron bacteria. *Caulobacter* (large "colony"), 1,000×, 102

F-3 Iron bacteria. *Thiobacillus ferro oxidans* (note precipitated iron), 1,000×, 103

F-4 Iron bacteria. *Pseudomonas,* 1,000×, 103

F-5 Sulfur bacteria. *Thiobacillus* (note precipitated sulfur), 1,000×, 104

F-6 Sulfur bacteria. *Beggiatoa* (note sulfur granules), 1,000×, 104

F-7 Sulfur bacteria. *Thiothrix* (note sulfur granules), 1,000×, 105

F-8 Sulfur bacteria. *Thiopedia* (note sulfur granules), 1,000×, 105

F-9 Nitrifying bacteria. *Nitrosomonas* (cocci), 1,000×, 106

F-10 Nitrifying bacteria. *Nitrobacter* (rods), 1,000×, 106

F-11 Nematode larva, 60×, 107

F-12 Nematode, 250×, 107

F-13 Chironomid. Head of a pupa, dorsal view, 1:7, 108

F-14 Chironomid. Pupa (Insecta, Chironomidae), ventral view, 1:20, 108

F-15 Chironomid. Pupa (Insecta, Chironomidae), ventral view, 1:70, 109

F-16 *Daphnia* (2 mm), 110

F-17 *Daphnia* sp. Water Flea (Crustacea), 110

F-18 *Cyclops* (2 mm), 111

F-19 *Cyclops* sp. Water Flea (Crustacea), 111

F-20 *Asellus aquaticus* (5 mm), 112

F-21 *Asellus aquaticus* (Crustacea, Isopoda), dorsal view, 1:5, 112

F-22 *Asellus aquaticus* (Crustacea, Isopoda), lateral view, 1:20, 113

F-23 *Asellus aquaticus* (Crustacea, Isopoda), head, dorsal view, 1:25, 113

F-24 *Asellus aquaticus* (Crustacea, Isopoda), head, ventral view, 1:100, 114

F-25 Faecal pellets of *Asellus aquaticus,* 1:4, 114

F-26 Faecal pellets of *Asellus aquaticus* containing bitumen, 1:40, 115

F-27 Faecal pellets of *Asellus aquaticus* containing iron-rust, 1:40, 115

F-28 Rotifer. *Monostyla,* 60×, 116

F-29 Rotifer. Gastrotrich (*Chaetonotus*), 200×, 116

F-30 Protozoan. Ciliate, *Vorticella,* unstained, bright-field microscopy, 120×, 117

F-31 Rotifers. Rotifer, 200×, 117

F-32 Gastrotrich, 200×, 118

F-33 Zebra mussels. Variations, 118

F-34 Zebra mussel larvae showing some disintegration, 119

F-35 Algae. Chlorophyte, *Scenedesmus,* unstained, bright-field microscopy, 120×, 119

F-36 Algae. Chrysophyte, *Dinobryon* (2 cells) and others, unstained, bright-field microscopy, 240×, 120

F-37 Algae. Diatom, *Asterionella,* unstained, bright-field microscopy, 120×, 120

F-38 Algae. Diatom, *Fragilaria,* unstained, bright-field microscopy, 240×, 121

F-39 Algae. Diatom, *Hannaea,* unstained, bright-field microscopy, 240×, 121

F-40 Algae. Euglenophyte, *Phacus,* unstained, bright-field microscopy, 120×, 122

F-41 Algae. *Anabaena Scherenietieri*, bright-field microscopy, 400×, 122

F-42 Protozoa. Amoeba, 200×, 123

F-43 Protozoa. Flagellates, 1,000×, 123

F-44 Protozoa. Free ciliate, 200×, 124

F-45 Protozoa. Stalked ciliate (stalks), 200×, 124

F-46 Protozoa. *Giardia lamblia,* 240×, 125

F-47 Protozoa. *Cryptosporidium*, 125

F-48 *Cryptosporidium* oocyst exhibiting sporozoites, DIC microscopy, 1,000×, 126

F-49 *Cryptosporidium* oocyst exhibiting internal structures known as sporozoites, phase contrast, 1,000×, 126

F-50 *Giardia* cyst and *Cryptosporidium* oocyst, stained with fluorescent antibody stain, 1,000×, 127

F-51 *Giardia* cyst exhibiting internal structures—nuclei, axonemes, median bodies, DIC microscopy, 1,000×, 127

F-52 Taste-and-odor algae, 128

F-53 Filter-clogging algae, 129

F-54 Polluted-water algae, 130

F-55 Clean-water algae, 131

F-56 Plankton and surface algae, 132

F-57 Reservoir algae, 133

F-58 Wastewater-treatment-pond algae, 134

F-59 Estuarine pollution algae, 135

Tables

11-1 Taxonomic listing of protozoa potentially associated with potable water, 73

B-1 Troubleshooting guide for problem organisms, 88

C-1 Suggestions for optimizing conventional water treatment systems, 92

This page intentionally blank.

Foreword

The drinking water community shares its water systems with a world of microscopic living creatures. These creatures may contribute to numerous operational challenges, from problems in source water through finished water, including encrustation or corrosion within the water system and water quality problems at the customer's tap. When water treatment problems arise due to these organisms, it is often difficult for an operator to locate help. This manual provides information regarding nuisance organisms, the problems they cause, treatment options, and references to literature resources. It is intended to be a "first stop" toward finding answers.

This third edition replaces AWWA Manual M7, *Problem Organisms in Water: Identification and Treatment,* 2nd edition. This complete revision of AWWA Manual M7 stems from the results of the 1989 AWWA Survey of Nuisance Organisms (appendix A) that demonstrated a need for information to help identify and control these organisms.

This page intentionally blank.

Acknowledgments

The AWWA Organisms in Water Committee thanks the following authors for sharing their knowledge and time:

Chapter 1: *John Lisle,* University of South Florida, Tampa, Fla., and *Bruce MacLeod* and *Mark Simpson*, Manatee County Public Works Department, Bradenton, Fla.
Chapter 2: *Jennifer L. Clancy,* Clancy Environmental Consultants, St. Albans, N.Y., and *Stuart A. Smith,* Ground Water Science, Upper Sandusky, Ohio
Chapter 3: *D. Roy Cullimore,* Regina Water Research Institute, Regina, Sask.
Chapter 4: *Nancy I. Lieu* and *Roy L. Wolfe,* Metropolitan Water District of Southern California, La Verne, Calif.
Chapter 5: *Richard C. Lorenz,* City of Westerville, Westerville, Ohio
Chapter 6: *Jeffrey Hoshaw,* Peerless-Midwest Inc., Mishawaka, Ind.
Chapter 7: *Eugene W. Rice,* Risk Reduction Engineering Laboratory, USEPA, Cincinnati, Ohio
Chapter 8: *Glenn Matherene,* Terre Bonne Parish Waterworks, Houma, La.
Chapter 9: *James F. Lubner,* University of Wisconsin Sea Grant Institute, Milwaukee, Wis.
Chapter 10: *George Izaguirre,* Metropolitan Water District of Southern California, La Verne, Calif.
Chapter 11: *Frank W. Schaefer III,* Risk Reduction Engineering Laboratory, USEPA, Cincinnati, Ohio, and *Susan N. Boutros,* Environmental Associates Ltd., Ithaca, N.Y.

Thanks also to Angela Podesta for organizing the Nuisance Organisms survey and the chapters in the manual. The committee hopes that this manual will be a useful resource for the water industry.

Organisms in Water Committee

M.M. Marshall, Chair, University of Arizona, Tucson, Ariz.
R.M. Hoffman, Vice-Chair, Wisconsin State Laboratory of Hygiene, Madison, Wis.
Zia Bukhari, American Water, Belleville, Ill.
Kevin Connell, DynCorp I&ET, Alexandria, Va.
Ricardo DeLeon, Metropolitan Water District of Southern California, La Verne, Calif.
G.D. Di Giovanni, Texas A&M University System, El Paso, Texas
C.R. Fricker, CRF Consulting, United Kingdom
N.H. Hall, University Hygienic Laboratory, Iowa City, Iowa
P.T. Klonicki, CH Diagnostic & Consulting Services Inc., Loveland, Colo.
A.B. Margolin, University of New Hampshire, Durham, N.H.
W.J. Robertson, Health Canada, Ottawa, Ont.
P.A. Rochelle, Metropolitan Water District of Southern California, La Verne, Calif.
F.W. Schaefer, III, USEPA, Cincinnati, Ohio
D.M. Sharp, Tampa Water Department, Tampa, Fla.
J.L. Sharpley, HDR Engineering, Bellevue, Wash.
Michelle Smith, Baldwin City, Kan.
Jon Standridge, Wisconsin State Laboratory of Hygiene, Madison, Wis.
Greg Sturbaum, CH Diagnostic & Consulting Service, Inc., Loveland, Colo.
Giovanni Widmer, Tufts University, North Grafton, Mass.
J.T. Wierenga, City of Grand Rapids, West Olive, Mich.

This page intentionally blank.

Introduction

In 1989 the Organisms in Water Committee of the American Water Works Association (AWWA) conducted a survey to determine problems that water utilities encounter due to organisms in water. Of the 80 systems surveyed, only 7 indicated that they experienced no problems due to organisms in water. The other 73 systems had their hands full. These problems went beyond finding low levels of coliform bacteria in routine monitoring samples. The difficulties were related to "nuisance organisms," microorganisms that cause problems with plant operations or with customers because of undesirable tastes, odors, color, or their mere presence in what is supposedly a clean glass of water.

NUISANCE ORGANISMS SURVEY

Appendix A presents the results of the Nuisance Organisms survey. Algal cells were identified as the greatest problem by 60 percent of the respondents. Blue–green algae causing taste-and-odor problems generated the majority of these complaints.

Tied for the second-most prevalent problem was iron bacteria within the distribution system or at the wellhead. These bacteria caused consumer taste-and-odor complaints. More than half the respondents sought help from a wide variety of sources, including other utilities, state or US Environmental Protection Agency (USEPA) laboratories, private consultants, laboratories, and universities. Typical control measures, such as increased chlorination, were only marginally effective.

Another common problem for utilities were actinomycetes, which cause particularly difficult taste-and-odor problems. Activated carbon and copper sulfate were found to be more effective controls than increased chlorination.

Midge larvae and other types of larvae were problematic, because it is the customer who usually discovers the problem. Increased chlorination and filter backwash were marginally effective, but utilities found no "sure-fire" approaches to this challenge.

Sulfur bacteria presented problems for 9 percent of the respondents. Because humans are particularly sensitive to hydrogen sulfide, small amounts of sulfur bacteria can cause offensive taste-and-odor problems.

The following chapters provide basic information and resources to assist the water treatment operator in troubleshooting and resolving problems caused by nuisance organisms. When problems arise, there is often no place for an operator to turn for help. This manual provides a resource to find more information regarding nuisance organisms in water. Each section provides a brief description of the organism, the problems that it causes, information on how to treat the problem, and references to recent literature to aid the operator in solving water treatment problems caused by organisms. Guides for troubleshooting the organisms discussed in this manual and for optimizing conventional water treatment systems may be found in appendixes B and C.

This page intentionally blank.

Chapter 1

Actinomycetes

Actinomycetes are a collection of nine different groups of bacteria. The most familiar of the group are the Streptomycetes. These organisms are widely distributed in nature and account for a large part of the normal microbial population of soils and water sediments.

Actinomycetes' significance to the potable water industry relates to their ability to produce, as part of their normal metabolic activities, earthy–musty–moldy taste-and-odor compounds. Five taste-and-odor compounds have been attributed to the actinomycete group. These compounds are

- geosmin or *trans*-dimethyl-*trans*-9-decalol, which imparts an earthy–musty taste and odor
- 2-methylisoborneol (MIB), which imparts an earthy–musty, camphor-like taste and odor
- 2-isopropyl-3-methoxy/pyrazine, which is similar to MIB
- 2-isobutyl-3-methoxy/pyrazine, which has an earthy–musty taste and odor sometimes described as similar to bell peppers
- 2,3,6-trichloroanisole, which is described as musty

BIOLOGY AND ECOLOGY

Actinomycete strains include three types—anaerobic, aerobic, and microaerophilic (requiring a small amount of oxygen to live). This diversity is an advantage in regard to their survival and reproduction in soils and water sediments. They have two different cell forms, filamentous and spore. The most frequently found form is determined by the growth conditions in their immediate environment. The filamentous or vegetative stage is the attached life stage, shown in Figure 1-1, during which they reproduce, producing taste-and-odor compounds.

The alternative stage is the spore form. Although they do not produce taste-and-odor compounds during this stage of the life cycle, it is in this resistant stage that they are easily dispersed in air, land, and water.

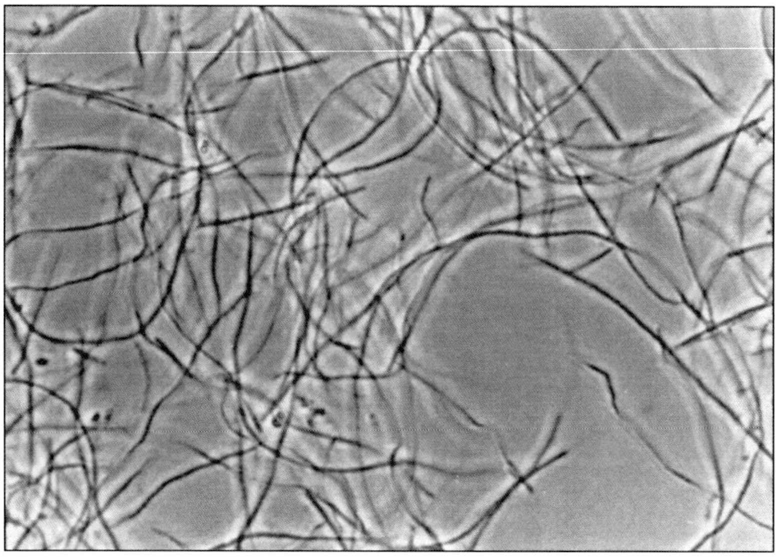

Source: Scott Tighe, Analytical Services Inc.

Figure 1-1 Actinomycetes, phase contrast 1,000×

When nutrition and environmental conditions are favorable, spores can germinate into the vegetative stage and produce taste-and-odor compounds. Studies indicate that the vegetative stage grows best in the temperature range of 15 to 38°C. This observation undoubtedly explains why taste-and-odor problems generally occur in the spring, summer, and fall months in temperate areas.

Actinomycetes have been isolated from various aquatic habitats. In general, these organisms prefer the following conditions:

- eutrophic ponds and lakes rich in nutrients, with low concentrations of dissolved oxygen
- shallow ponds
- both clear and turbid lakes
- sediments, although actinomycetes can be found throughout the water

Identification

The transparent body of the vegetative actinomycetes that appears in Figure 1-2 makes these organisms difficult to observe with a compound microscope.

Actinomycetes' general size and shape can be easily confused with other organisms that are not involved in taste-and-odor production. The spore stage, which is even more difficult to observe because of its very small size, is visible at higher magnifications with the light microscope.

Shown on the left in Figure 1-3 is a typical bacterial colony characterized by a smooth mucous-like appearance and a smooth border. On the right is an actinomycete colony characterized by branching filaments that give it a fuzzy appearance at its border and a dull powdery appearance.

Direct examination of soils, water, and sludges. Actinomycetes are most easily identified by culturing and microscopic examination. Because this process requires from four to seven days to complete, it is not practical for immediate operational control. The customer will be aware of the taste and odor long before the cultures

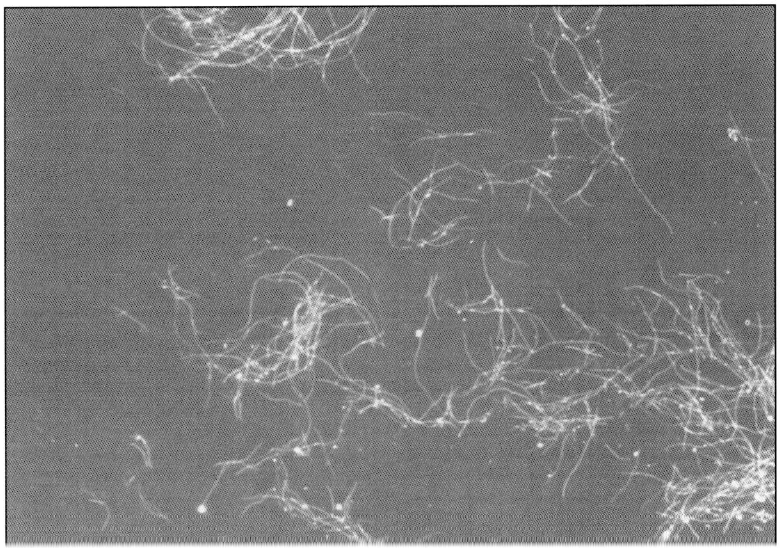

Source: Scott Tighe, Analytical Services Inc.

Figure 1-2 Actinomycetes, dark-field illumination, 400×

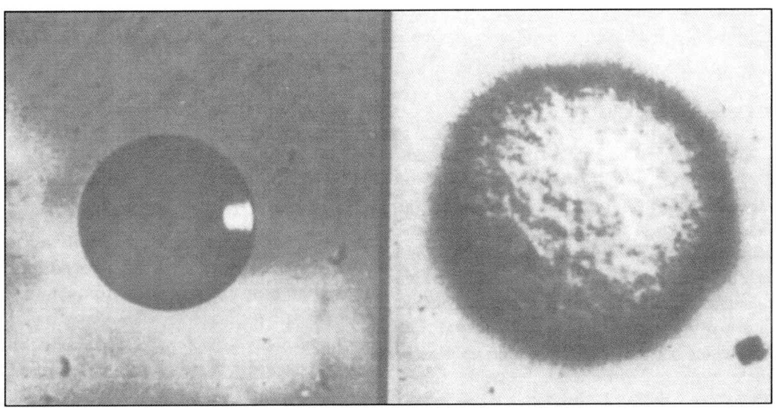

Source: Standard Methods for the Examination of Water and Wastewater.

Figure 1-3 Bacterial colonies—typical colony versus actinomycete colony, 50×

are ready for examination. However, regular sampling and examination for actinomycetes will enable plant staff to develop baseline data for a source water. With these data, staff may predict when problems will occur and make operational changes before the organism becomes a problem.

Cellular morphology. The vegetative stage is characterized by branching and filamentous cells. The filaments can be extremely short, making them hard to distinguish from nonfilamentous, nonactinomycete organisms. Individual filaments vary in diameter from 0.5 to 2.0 µm. Spores are most easily observed when they are attached to filamentous branches. If the spore chains do not break, they may appear as a straight or looped chain. A simple stain, such as crystal violet or safranin, will increase contrast to view specimens more easily under the microscope.

Culture characteristics. The most productive method of observing actinomycetes is to grow them in isolated colonies on a specific agar. This method is more efficient than isolating a single vegetative cell from a reservoir. Samples take from four to seven days to incubate before microscopic examination. Five media that can be used to isolate actinomycetes are listed in appendix D.

Indirect methods of examination. Odor and flavor testing are useful to determine the concentration of earthy or musty odors that accompany the presence of actinomycetes. There are currently four methods used to determine the concentrations of odors produced by actinomycetes, including threshold odor, flavor profile analysis, gas chromatography, and gas chromatography–mass spectroscopy. The first two tests, threshold odor and flavor profile analysis, are sensory. They are qualitative tests but have the advantage of being quick, inexpensive, and require little training to perform. The threshold odor test is often sufficient for MIB detection, because that compound has a strong smell.

The flavor profile analysis test can be used only on samples safe for consumption. These methods are described in detail in *Standard Methods for the Examination of Water and Wastewater,* Sections 2160 and 2170. The flavor profile method is more quantitative than the threshold odor test. It requires an intermediate level of training and expertise and a panel of at least four trained members to describe the intensities and characteristics of tastes and odors. Reference samples of earthy–musty compounds are required, as well as periodic calibration of the panelists. The panel determines the concentration based on the flavor intensities of the samples.

Two other tests—gas chromatography or gas chromatography–mass spectroscopy—can also be used to measure actinomycete-caused taste-and-odor compounds. The ability of humans to detect these tastes and odors at low levels, however, requires a concentration step before analysis. These methods require expensive equipment and highly trained analysts to yield quantitative information down to 5.0 parts per trillion. These methods are described in detail in *Standard Methods for the Examination of Water and Wastewater,* Section 9250.

SIGNIFICANCE FOR WATER SUPPLIES

Actinomycetes have been isolated from soils, source water, and drinking water treatment plants. The presence of these organisms in soil may become a problem for water utilities when organisms are carried in runoff into source water supplies, establish themselves, and begin producing odors.

It is difficult to positively correlate actinomycete numbers with a taste-and-odor problem since some actinomycetes' spore colonies do not produce tastes and odors.

Actinomycete spores or vegetative cells in the source water may be drawn into a water plant's intake. They can associate with floc and settle in the sludge blanket or in algal mats along the walls of treatment basins. If growth conditions are favorable, actinomycetes may release taste-and-odor compounds. These compounds may also be introduced into the potable water supply when the following conditions occur:

- organisms or spores are retained on treatment plant filters
- soil enters the distribution system during installation of water pipes and repair of main breaks
- a cross connection occurs with an actinomycetes-contaminated source, such as an irrigation pond system connected to standard plumbing without check valves

These taste-and-odor compounds can be detected by humans at extremely low concentrations, for example, 2.0 parts per trillion for MIB. Although consumption of these bacteria or their compounds is not considered a health hazard, the water may still be aesthetically unacceptable to the consumer.

CONTROL STRATEGIES

Water treatment techniques for controlling taste-and-odor compounds produced by actinomycetes depend on the source of the difficulty. Distribution system problems are generally easier to correct than problems in the source water or treatment plant.

Flushing is an effective course of action when the problem is in an isolated area of the distribution system. Actinomycete filaments and spores are susceptible to chlorination. If the problem is persistent, the operator should consider isolating the line and superchlorinating. Both pigging and superchlorination may be necessary if actinomycetes are associated with biofilms.

Taste-and-odor problems due to actinomycete growth at the treatment plant can be controlled by minimizing sludge depth in the sedimentation basin and by maintaining an active basin-cleaning program. Taste-and-odor problems in the source water are the most difficult to control. An alternate source that is taste-and-odor free should be considered first. Sometimes varying the intake depth can be useful.

The oxidants chlorine, chloramines, chlorine dioxide, and potassium permanganate are generally not effective in reducing taste and odor compounds caused by actinomycetes. Although some utilities have reported success using these oxidants, others have found their use increases the intensities of tastes and odors.

Ozone, and recently, hydrogen peroxide and ozone, called PEROXONE, have been effective in reducing the concentration of these odorants. Capital costs for this treatment technique are expensive and must be justified on a case-by-case basis.

Activated carbon adsorption has proven effective in many cases. Powdered activated carbon can be added during taste-and-odor episodes to the raw water at an application point that maximizes contact time. Granular activated carbon has been used successfully either as a filter medium or as a postfilter contactor. When considering carbon, it is important to minimize other organics that may interfere with the adsorption of the target odorants and maximize activated carbon and odorant contact time.

REFERENCES

American Public Health Association, American Water Works Association, Water Pollution Control Federation. 1992. *Standard Methods for the Examination of Water and Wastewater.* Latest edition. Washington, D.C.: American Public Health Association.

Arai, Tadashi. 1976. *Actinomycetes—The Boundary Microorganisms.* Tokyo, Japan: Toppan Company.

Bartel, J.H.M., B.M. Brady, and I.H. Suffet. 1989. *Taste and Odor in Drinking Water—Phase I and II.* Denver, Colo.: American Water Works Association Research Foundation and American Water Works Association.

Burlingame, G.A., and M.J. McGuire. 1994. Simple Test Flags Changes in Taste and Odor. *Opflow,* 20(10):1,6.

Difco Laboratories Inc. 1984. *Difco Manual.* 10th ed. Detroit, Mich.

Faust, S.D., and O.M. Aly. 1983. *Chemistry of Water Treatment.* Boston, Mass.: Ann Arbor Science Books.

Goodfellow, M., M. Mordarski, and S.T. Williams. 1983. *The Biology of Actinomycetes.* London, England: Academic Press.

Goodfellow, M., S.T. Williams, and M. Mordarski. 1988. *Actinomycetes in Biotechnology.* San Diego, Calif.: Academic Press.

Mallevialle, J., and I.H. Suffet (eds.). 1987. *Identification and Treatment of Tastes and Odors in Drinking Water.* Denver, Colo.: American Water Works Association Research Foundation and American Water Works Association.

Silvey, J.K.G. 1988. Tastes and Odors. In *Manual of Water Utility Operations.* 8th ed. C.K. Foster, ed. Austin, Texas: Texas Water Utilities Association.

Waksman, S.A. 1967. *The Actinomycetes: A Summary of Current Knowledge.* New York, N.Y.: Ronald Press.

This page intentionally blank.

Chapter 2

Iron Bacteria

Iron-related or iron-precipitating bacteria (iron bacteria) are a diverse group of microorganisms widely distributed in nature. They are found in fresh and salt waters, in soils, and on desert rock surfaces. Iron bacteria do not normally cause diseases to humans or animals, but they are nuisance microorganisms. These bacteria are capable of transforming iron and sometimes manganese to an insoluble form that can cause severe fouling problems in wells, treatment plants, and distribution systems. Similar problems are caused by bacterial formation of sulfur slimes, which are formed by a separate group of bacteria (see chapter 3). Iron bacteria can also cause fouling in industrial cooling and boiler waters.

BIOLOGY AND ECOLOGY

Iron bacteria convert soluble iron (Fe^{2+}), either naturally present in water or corroded from pipe surfaces, to the insoluble form (ferric iron [Fe^{3+}]), which is deposited on or outside the bacterial cells. In combination with bacterial polymers, these insoluble forms are responsible for fouling. Iron bacteria are able to accumulate large amounts of this material, well in excess of their biomass. As can be seen in Figure 2-1, cells are surrounded by a deposit of ferric hydrate. On average, individual cells are 0.4 to 1.5 µm wide and 0.8 to 2.5 µm long.

Bacteria that can be associated with iron biofouling include the following:

1. The sheathed bacteria, such as *Leptothrix* and *Clonothrix*. These produce sheaths of oxidized iron that surround their cells. Oxidized iron or manganese is deposited outside the sheaths. *Sphaerotilus*, a form that may be related to *Leptothrix*-type bacteria, is familiar as "sewage fungus" and can form large flocs that may interfere with activated sludge processes.
2. Bacteria with appendages, such as *Hyphomicrobium*, *Caulobacter*, and *Gallionella*. *Gallionella* form long stalks on which iron is deposited outside the cells. The bacteria *Hyphomicrobium* and *Caulobacter* are associated with manganese precipitation.
3. *Thiobacillus* bacteria can thrive in aerobic or anaerobic environments and can oxidize both reduced sulfur and reduced or ferrous iron. These are the only known iron-oxidizing bacteria and are considered to be both iron and sulfur bacteria.

Source: Standard Methods for the Examination of Water and Wastewater.

Figure 2-1 Single-celled iron bacterium *Siderocapsa treubii*

4. Heterotrophic iron-precipitators in the environment precipitate iron mostly by using the organic part of iron-organic complexes and discarding the iron. This group includes a variety of coliforms and members of *Pseudomonas* and related bacteria.

Iron bacteria are common in water. The *Sphaerotilus–Leptothrix* group is thought to proliferate under low organic nutrient conditions in flowing water. They attach to surfaces in biofilms and thrive because nutrients are constantly provided and waste products are washed away. Figure 2-2 shows *Sphaerotilus natans* with cells within filaments and some free "swarmer" cells. Filaments show false branching and some areas without cells. Individual cells within the sheath may vary in size, averaging 0.6 to 2.4 µm wide by 1.0 to 12.0 µm long. Most strains are 1.1 to 1.6 µm wide and 2.0 to 4.0 µm long.

Lakes rich in suspended solids may contain high concentrations of *Gallionella*, *Hyphomicrobium*, and *Clonothrix*. Such bacteria are present in almost all aquifers studied worldwide. Examples of iron bacteria, including *Leptothrix*, *Caulobacter*, *Thiobacillus*, and *Pseudomonas,* appear in the color section of this manual.

Figure 2-3 illustrates a laboratory culture of *Gallionella ferruginea*, showing cells, stalks excreted by cells, and branching stalks where cells have divided. Inorganic iron on and around the stalks often blurs the outlines. Cells at the tip of the stalk average 0.4 to 0.6 µm wide by 0.7 to 1.1 µm long.

Environmental factors including temperature, light, pH, redox potential (Eh), organic matter, and iron or manganese content determine whether or not these bacteria will thrive and proliferate to an extent that results in fouling and, thus, requires treatment. Figure 2-4 shows a mixture of fragments of stalks of *Gallionella ferruginea* and inorganic iron–manganese precipitate found in natural samples from wells. Stalks will appear golden yellow to orange when examined under the microscope.

Soluble inorganic and organic material in groundwater can be oxidized microbially to a soluble state when exposed to oxygen. Sheaths and slimes may be produced that can impart color, tastes, and odors to water. The sheaths and slime can become encrusted with iron; plug sand filters, pump impellers, and column pipe; and interfere with filtration and distribution of drinking water.

Identification

Culture methods for a number of types of iron-oxidizing bacteria can be found in *Standard Methods for the Examination of Water and Wastewater,* Section 9240. Culture methods are useful for two reasons. They provide an "early warning" of iron-precipitating

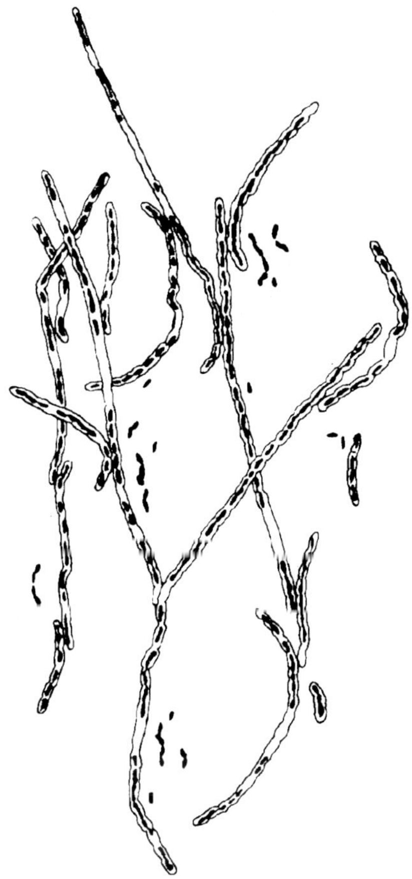

Source: Standard Methods for the Examination of Water and Wastewater.

Figure 2-2 Filaments of *Sphaerotilus natans*

Source: Standard Methods for the Examination of Water and Wastewater.

Figure 2-3 Laboratory culture of *Gallionella ferruginea*

Source: Standard Methods for the Examination of Water and Wastewater.

Figure 2-4 Stalk fragments of *Gallionella ferruginea* and iron–manganese precipitate

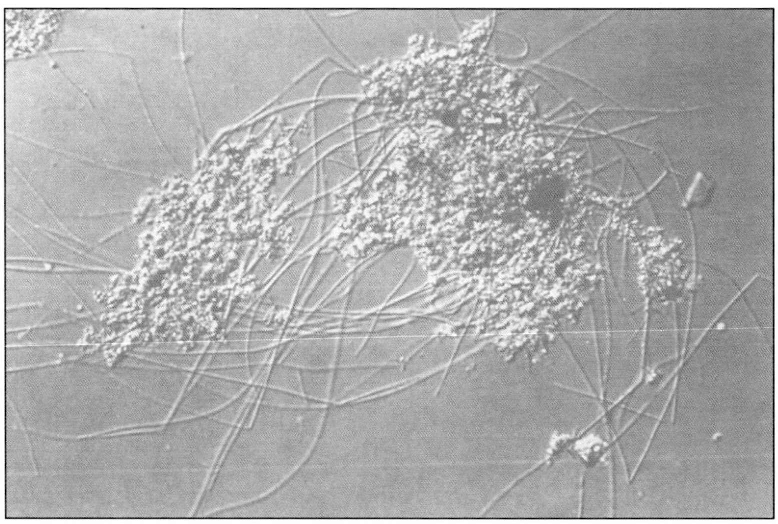

Source: Scott Tighe, Analytical Services Inc.

Figure 2-5 Filamentous iron bacteria recovered from well water. DIC microscopy, 1,000×

heterotrophic bacteria, which can easily be missed by microscopic methods. They also can be used to confirm the presence of living bacteria in iron deposits.

Direct examination. When iron bacteria have become so numerous that they affect water quality, culture methods are rarely needed. Microscopic examination of the slime, floc, or other masses of microbial material is usually all that is required to determine whether or not iron bacteria (at least filamentous or stalked types) are the problem (see Figure 2-5). Routine cultural examinations provide confirmation that the iron precipitation is "alive." Early identification enables plant staff to change operational practices before the iron bacteria form intractable deposits.

Source: Stuart Smith.

Figure 2-6 Partly filled impeller channel

Microscopic methods to determine presence or absence and quantity of iron bacteria in a sample are taken from *Standard Methods for the Examination of Water and Wastewater*, Section 9240, and may be found in appendix E.

SIGNIFICANCE FOR WATER SUPPLIES

Bacterial iron precipitation, or iron biofouling, can cause taste-and-odor problems, frothing, color, and increases in turbidity. Iron bacteria can mask the presence of pathogenic bacteria, reduce the efficiency of disinfection, and increase disinfectant demand. Biofilms surrounding iron bacteria harbor corrosion-producing bacteria and accelerate corrosion. Once established, biofilms are often unaffected by the chlorine levels usually found in drinking water distribution systems.

Iron bacteria can multiply rapidly in distribution piping under favorable conditions. Accumulations of insoluble iron minerals and biofouling on the pipes themselves reduce flow.

The sequence shown in Figures 2-6 through 2-9 demonstrate iron biofouling in a well pump (lineshaft turbine) in western Ohio. The figures show iron biofouling typically occurring over 2 to 10 years in this area. Yield had declined from over 380 gpm to less than 200 gpm over 10 years. Corrosion of castings under the iron biofilm weakened this pump beyond repair. The encrusted material is mostly ferrihydrite, but there are some filamentous forms.

These chemical and physical changes on the pipe surfaces can lead to the growth of other types of microorganisms that can cause further taste, odor, and corrosion problems.

Iron bacterial fouling deteriorates piping in distribution lines, reduces distribution capacities, and increases operating and maintenance costs. Iron biofouling especially affects well pumps and column pipe, causing plugging, wear, and corrosion. Inorganic iron deposits and organic slimes produced by these bacteria clog pump intakes, well screens, pipelines, and filters.

Severe fouling in private wells has ruined washers, dishwashers, toilets, tubs, and clothing. In addition, iron biofouling allows protected anaerobic environments to form, promoting corrosion, forming hydrogen sulfide and black ferrous sulfide deposits.

Source: Stuart Smith.

Figure 2-7 Largely clogged bowl volutes

Source: Stuart Smith.

Figure 2-8 Column pipe with iron biofilm

CONTROL STRATEGIES

Controlling iron bacteria in a water system may be achieved either by thorough removal or inactivation of the bacteria in the system or by removal of the bacteria from the water.

Preventing Iron Bacteria Fouling in Wells

The major factors creating iron bacteria fouling include the following:

- the presence of the bacteria
- the presence of dissolved or complexed iron or manganese
- an environment that encourages bacterial survival and growth

Source: Stuart Smith.

Figure 2-9 Clean versus clogged impellers

Aggravating factors are controllable by sound engineering, construction, and operating practices. Controlling the following factors will limit the impact of iron bacteria on a well and its connected system:

- well design, construction, and use
- water system design and construction
- choices in water treatment

Design or construction flaws allow the entry of bacteria, increase chemical oxidation, restrict inlet channels, and provide havens for growth. Cascading water from breaks or fractures above the pumping water level encourages heavy biofouling growth and pump plugging.

Periods of nonuse or occasional use allow fouling growths to increase. However, in some cases, intermittent use may dissipate heavy biofouling. Overuse may aggravate fouling buildup by encouraging sand or mineral clogging and extra oxidation.

The use of corrodible metals in the well and pumping system, bimetallic connections (bronze/cast impeller assemblies), and the mixing of waters of differing redox potential encourage corrosion supported by bacterial activity.

The key to catching growth before it causes problems is ongoing, preventive monitoring and testing. Official "standard" methods to detect either bacteria involved in biofouling or the biofouling itself do not yet exist. However, the cost-effective methods described earlier enable identification of bacteria before symptoms occur. Hydraulic testing (drawdown and yield) should be conducted periodically to check for reduced function. Turbidity testing can also detect bacterial particle breakthrough and indicate problems in the system.

Tests should be run soon after a well is drilled or serviced in any way, and also at regular intervals—monthly, quarterly, or annually. On highly valuable wells, borehole camera surveys are useful in monitoring the progress of biofilm buildup in the well.

Valuable wells should be tested hydraulically using step-test procedures one to four times a year. Such wells should be equipped for convenient flow, line pressure, and drawdown measurement. Valved diversions allow pumping tests to be conducted without concern for varying system total dynamic head (TDH). Measurement apparatus should be kept in good working order by cleaning such equipment as flowmeters and air lines. Step tests allow an analysis of changes in specific capacity and total yield and can be used to distinguish pump from aquifer problems. If caught early, regular chlorination and increased vigilance can keep bacterial problems under control.

It is important to avoid introducing iron bacteria into a well during drilling or repair work, although the most likely bacteria source is the aquifer itself. Chlorinate all drilling fluid mix, makeup, and mist water, even if it is from a chlorinated municipal supply. Because of chlorine's instability, add it periodically to maintain an appropriate free-chlorine residual. Drilling-mud manufacturers have recommendations on maximum chlorine levels. The removal of phosphate and organic polymer additives should be confirmed by tests during well development.

To avoid contamination from one well to another, disinfect drilling rods, bits, and tools thoroughly with chlorine between jobs. Chlorinate pumps, pipe, and filter-pack material before installation. Carefully washing pack material with chlorinated water ensures it is free of organic matter. The well should be chlorinated (refer to ANSI/AWWA C654, Standard for Disinfection of Wells, latest edition) and sealed immediately to prevent the introduction of airborne contamination, unless it is tested and put into service immediately after construction or repair.

Controlling Iron Bacteria Fouling in Wells

Once iron bacteria become established in a well, chemical and physical treatment may be required to control bacterial problems. Shock chlorination of the well may eliminate the problem for one to three years, after which the procedure will need to be repeated. Once established, such treatments tend to be necessary at increasingly shorter intervals, decreasing to monthly or weekly.

In some systems, continuous chlorination may be required to keep the iron bacteria from proliferating. Chlorination followed by greensand or conventional filtration has been effective in a number of systems. Chlorine, hypochlorites, and potassium permanganate have been used for chemical control of iron bacteria. For maximum effectiveness, shock chlorination should be accompanied with physical agitation of the well. This may include jetting, air surging, air-lift pumping, or valved surge blocks. After shock chlorination, the well should be treated with a dissolution–dispersion step using acidification and surfactants, then surged. A maintenance contract between a competent well-cleaning contractor and the well owner is advisable.

Physical methods of removing iron bacteria from wells have included heat pasteurization, explosives, and ultrasound. The latter two methods have not been reported to be successful and should be avoided in most cases. High-frequency sonic methods, however, may be used to remove hard outer deposits when treating iron biofouling.

One method of chemical control involves injecting chlorine into ground wells drilled around supply wells. Another procedure involves the injection of oxygenated degassed air into the subsurface as a means of in situ biooxidation. This method encourages bacterial growth near the wells, theoretically reducing growth near the source water well. Increasing the redox potential of groundwater leading to precipitation of iron and manganese has been used in some sand and gravel aquifers to isolate iron in the groundwater before water reaches the well, but this treatment has had limited application.

Further information on control of well iron biofouling can be found in *Evaluation and Restoration of Water Supply Wells* (1993), published by AWWA and the AWWA Research Foundation. Employ an experienced well-remediation contractor and include testing in contract specifications to apply remediation methods exactly.

Control of Iron Bacteria in Filters

Controlling iron bacteria growth in filters may be difficult, because the conditions permitting their proliferation (for example, nutrients in the source water, light, pH, and Eh of the water) may be beyond the operator's control. Once established, bacteria can be removed by backwashing and superchlorination. If the problem is short-lived or seasonal, increasing the chlorine level at the prechlorination step may be necessary to keep the bacteria in control. Additional treatment, such as greensand filtration, may be required on a long-term basis.

Control of Iron Bacteria in the Distribution System

In both groundwater and surface water systems, problems associated with iron bacteria may not appear until the water is in the distribution system. The quality of water leaving the treatment plant or wellhead may deteriorate at some point in the distribution system, and fouling occurs. Conditions in piping materials, both chemical and physical, can enhance the growth of a relatively low level of iron bacteria to the point that they become a nuisance.

The result may lead to corrosion and tuberculation of the pipes. When levels of iron and iron bacteria in the finished water are already low, immediate attention should be focused on the pipes. Severely corroded pipes may require replacement. Shock chlorination of the lines may reduce high levels of iron bacteria on pipes. For dead-end areas, looping water lines may be necessary to maintain adequate chlorine residual for bacterial growth control. Physical removal of corroded materials, followed by the use of a backwashable, corrosion-control chemical, and filtration to prevent or limit reseeding of iron bacteria, may be necessary to prevent future water quality declines because of iron bacteria.

Consistent monitoring is essential to prevent a full-blown resurgence of iron bacteria contamination. Bacteria tend to grow back. If regrowth persists, periodic chlorination should be instituted.

REFERENCES

American Public Health Association, American Water Works Association, Water Pollution Control Federation. 1992. *Standard Methods for the Examination of Water and Wastewater.* Latest edition. Washington, D.C.: American Public Health Association.

American Water Works Association. 1987. ANSI/AWWA Standard C654, *Disinfection of Wells.* Latest edition. Denver, Colo.: American Water Works Association.

Cliver, D., R. Newman, and J. Cotruvo, eds. 1987. Drinking Water Microbiology. *J. Environmental Pathology, Toxicology, and Oncology,* 7:5,6.

Cullimore, D.R. 1990. An Evaluation of the Risk of Microbial Clogging and Corrosion in Boreholes. In *Water Wells Monitoring, Maintenance, Rehabilitation*. P. Howsam, ed. pp. 25–34. London, England: E. & F.N. Spon.

Dondero, N. 1975. The Sphaerotilus–Leptothrix Group. *Ann. Rev. Microbiol.,* 29:407–428.

Driscoll, F.D., ed. 1986. Well Failure Caused by Iron Bacteria. In *Groundwater and Wells*. pp. 646–655. St. Paul, Minn.: Johnson Filtration Systems Inc.

Mansuy, N., and C. Nuzman. 1990. Well Problems Identification and Its Importance in Well Rehabilitation. P. Howsam, ed. In *Water Wells Monitoring, Maintenance, Rehabilitation*. pp. 87–99. London: E. & F.N. Spon.

Ridgway, H.F., and B.H. Olson. 1981. Scanning Electron Microscope Evidence for Bacterial Colonization of a Drinking-Water Distribution System. *Applied and Environmental Microbiology,* 41:274–287.

Smith, S. 1982. Culture Methods for the Enumeration of Iron Bacteria From Water Well Samples—A Critical Literature Review. *Ground Water,* 20:482–485.

Smith, S.A. 1992. *Methods for Monitoring Iron and Manganese Biofouling in Water Wells*. Denver, Colo.: American Water Works Association Research Foundation and American Water Works Association.

Smith, S.A., and M.S. Borch. 1993. *Evaluation and Restoration of Water Supply Wells*. Denver, Colo.: American Water Works Association Research Foundation and American Water Works Association.

Starkey, R.L. 1945. Transformations of Iron by Bacteria in Water. *Jour. AWWA*, 37:963–984.

Tuovinen, O.H., and E.L. Nurmiaho. 1979. Microscopic Examination of Bacteria in Fe(III)-oxide Deposited From Ground Water. *Microbial Ecology*, 5:57–66.

Chapter 3

Sulfur Bacteria

Sulfur bacteria are serious nuisance organisms in water because they can cause severe taste-and-odor problems as well as corrosion. As sulfur bacteria grow, they use large quantities of sulfates, other forms of oxidized sulfur, and/or elemental sulfur to generate hydrogen sulfide gas. These bacteria, referred to as sulfate-reducing bacteria (SRB), create smells like "rotten eggs," initiate corrosive processes in metal fittings, and react with dissolved metals, such as iron, to generate black deposits. Note that rotten-egg smells also are created by other bacteria, including many of the coliforms.

Usually the events described above mean that somewhere upstream major bacterial fouling has removed oxygen from the water and allowed these bacteria to dominate. Bacteria commonly vent hydrogen sulfide into waters when oxygen is absent and sufficient amounts of dissolved organic materials exist. Forcing oxygen into the water eliminates the SRB, since oxygen is toxic to their activities.

BIOLOGY AND ECOLOGY

SRB usually protect themselves by cohabiting slimes and tubercles with other slime-forming bacteria. A tubercle may appear as a slightly raised area on the metal surface, as shown in Figure 3-1.

Figure 3-2 illustrates a cross section of a slime, with the sulfate-reducing bacteria located at the metal surface and other bacteria overlaying the slime. This creates a bubble-like structure composed of hardened iron-rich plates overlaying an active slime formation. Figure 3-3 shows the cross section of a tubercle.

Rotten-egg smells will occur more commonly when a water system or well is not used for an extended period of time. In these cases, the oxygen in the water is used up by various bacteria. Once the oxygen is gone, the growth and activities of the SRB group become rampant, with intense production of the rotten-egg smell.

Once an SRB group establishes itself on a surface within a biological slime or tubercle, it begins to generate hydrogen sulfide gas. This gas can trigger a complex electrolytic corrosion process on some metallic surfaces. Corrosion begins with pitting and cavitation and terminates with perforation of the supporting structures, such as metal pipe walls, and system failure.

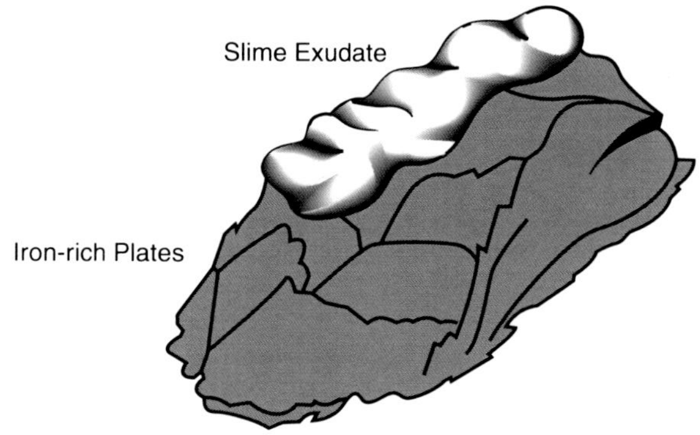

Source: D.R. Cullimore.

Figure 3-1 Side view of a tubercle

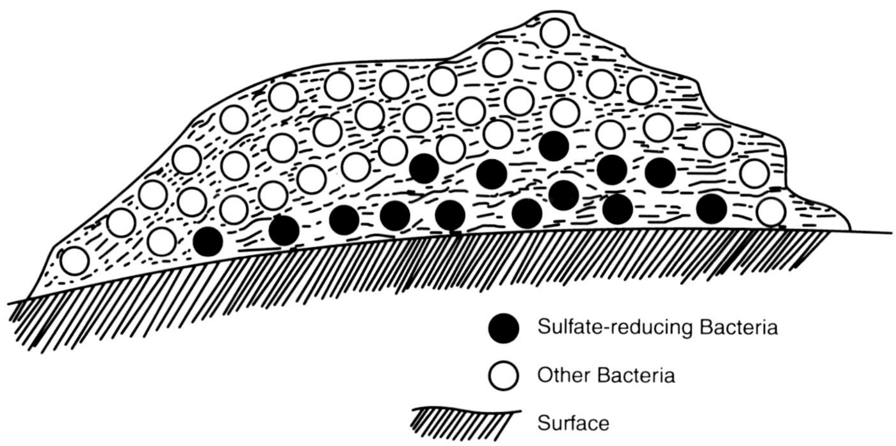

Source: D.R. Cullimore.

Figure 3-2 Cross section of a slime

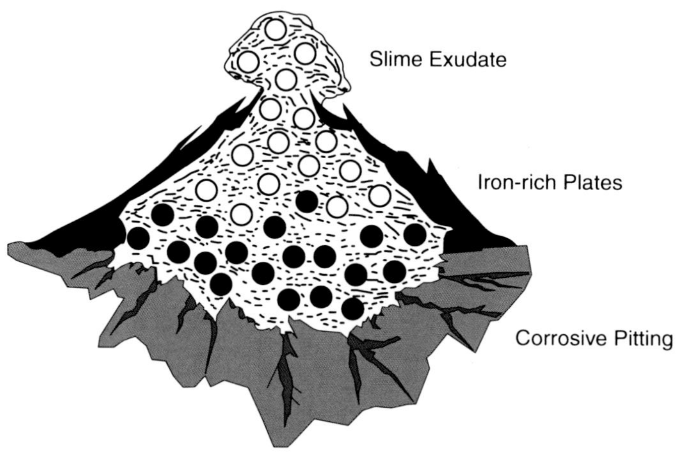

Source: D.R. Cullimore.

Figure 3-3 Cross section of a tubercle

At the same time, various slime-forming bacteria that inhabit the same sites may magnify the problem by generating various organic acids, which are also corrosive. These include sulfur-oxidizing bacteria, colorless sulfur bacteria, and purple and green sulfur bacteria. Examples of SRB *Thiobacillus*, *Beggiatoa*, *Thiothrix*, and *Thiopedia* are shown in the color section of this manual. Of these groups, the sulfur-oxidizing bacteria are the most documented because of their association with acidic mine tailings and leaching of ores. These bacteria require oxygen to grow and they convert various sulfides to acidic products, such as sulfuric acid.

The colorless sulfur bacteria do not usually produce acidic products but do convert hydrogen sulfide and other sulfides to sulfates. These bacteria are sometimes found growing in water wells and distribution systems where sulfides are present and where there is also a significant level of dissolved oxygen.

Oxygen is toxic to the purple or green sulfur bacteria which, like plants, are able to photosynthesize. Unlike plants, however, these bacteria oxidize sulfides to elemental sulfur, which becomes deposited in and around the cells. Common habitats for these bacteria include septic ponds, where they occasionally dominate and turn the water red, and deeper lakes that have become stratified, or layered, and where these bacteria form distinctive plates of floating growth.

Identification

Because the SRB cause corrosion while growing on a surface, tests on the water itself may be negative. Tests will be positive when bacteria are present in the water while moving from one corrosion site to colonize another.

The following is a taxonomic listing of the major genera of sulfur bacteria associated with potable water.

- Sulfate-Reducing Bacteria (SRB)
  Desulfovibrio
  Desulfotomaculum
- Sulfur-Reducing Bacteria (SRB, but use sulfur)
  Desulfuromonas
- Sulfur-Oxidizing Bacteria (produce acidic products)
  Thiobacillus
- Colorless Sulfur Bacteria (need hydrogen sulfide)
  Beggiatoa
  Thiothrix
- Sulfur-Oxidizing Photosynthetic Purple and Green Sulfur Bacteria
  Chlorobium
  Chromatium

For more detailed information on SRB, see *Standard Methods for the Examination of Water and Wastewater*, Section 9240. Figures 3-4 and 3-5 illustrate two sulfur-reducing bacteria.

Primary indicators of an SRB problem include the presence of odor, tubercles, corrosion inside metallic equipment, and black slimes. Bacteriological tests can provide confirmation of SRB involvement. The SRB grow in protected places, often surrounded by other types of bacteria that may mask their presence. These growths make microscopic examination difficult. Most culture methods, like those in Section 9240C of *Standard Methods for the Examination of Water and Wastewater*, confirm the presence

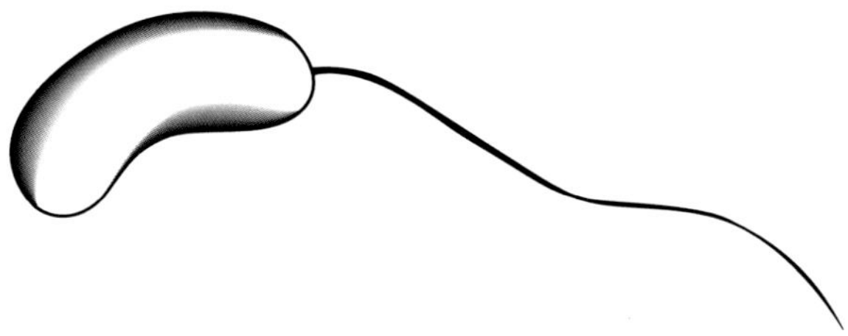

Source: D.R. Cullimore.

Figure 3-4 An SRB *Desulfovibrio desulfuricans*

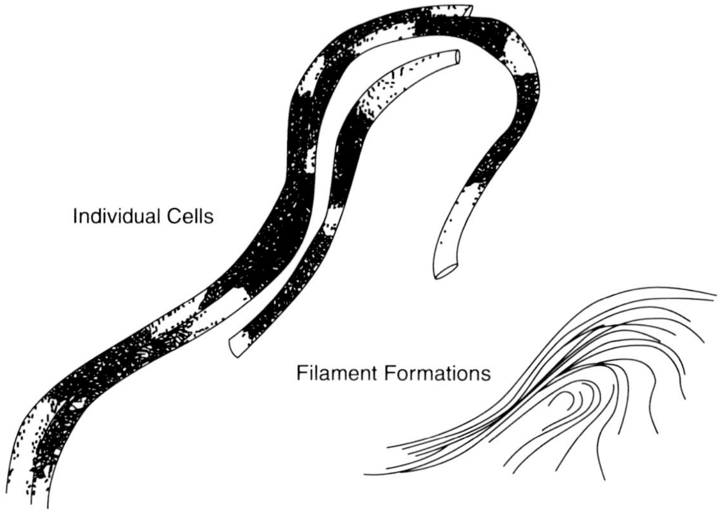

Source: D.R. Cullimore.

Figure 3-5 *Beggiatoa,* a colorless sulfur bacterium

of SRB when the bacteria produce visible black sulfides. Section 9240C of *Standard Methods for the Examination of Water and Wastewater* provides information on media to grow the more common types of SRB. These black deposits appear in the liquid culture medium or as distinctive black growths in or around SRB colonies. The lag time before these black deposits become visible indicates the aggressivity of the SRB group. For example, a rapid two-day delay would indicate a more aggressive population than an eight-day delay period.

SIGNIFICANCE FOR WATER SUPPLIES

An active SRB infestation may present multiple difficulties, including corrosion; severe taste, odor, and colored water problems; loss of efficiency or equipment failures in treatment processes; and increasing consumer complaints.

Corrosion is difficult to control once SRB become established. Slimes or tubercles inhibit chemical treatments using chlorine, acids, and cleaning agents. As a result, higher dosages and longer exposure times become necessary to control corrosion.

Where significant amounts of iron, manganese, or other metallic materials occur in the water, hydrogen sulfide can react with these compounds to form metallic sulfides. Many of these chemical compounds are black in color and can cause the slime to become black in appearance. When these blackened slimes break up, the water may

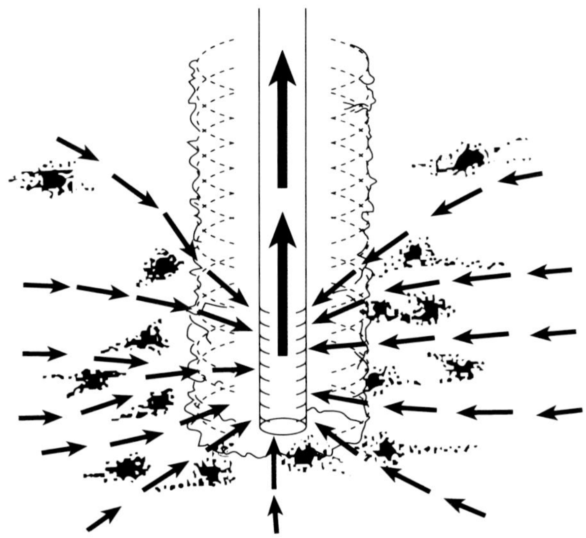

Source: D.R. Cullimore.

Figure 3-6 An operating water well being pumped, influence (suppressive) on SRB activity (theoretical)

contain threadlike strings of black slime-like material. These black growths frequently are not accompanied by rotten-egg smells, because the hydrogen sulfide gas has been converted into black sulfides. The slimes flow down walls and across surfaces and may change in color to browns, reds, yellows, and greys as the oxygen in the air stimulates aerobic bacterial activity in the slime.

CONTROL STRATEGIES

Unfortunately, the ability of SRB to grow in places where they are protected by either copious "overburdens" of slime or within tubercles makes control difficult. Treatments that have been recommended include a variety of disinfection, acidification, and cleaning practices. It is essential when applying these various practices to follow the suggestions below.

Step 1 Ensure that the system has been flushed and cleaned thoroughly before starting the treatment program.

Step 2 Apply the highest recommended dosage and use the longest contact time for the selected treatment program in order to maximize effectiveness.

Step 3 Increase the dissolved oxygen in the water to suppress SRB activity.

Step 4 Consider ongoing or routine application of disinfectants and penetrants to reduce the rate of recovery of the SRB from the "shock" effect of step 2.

Pump operation influences the effect of SRB, as the increased flow to an operating pump will tend to discourage SRB growth. A quiescent or inactive pump causes the spread of SRB, as oxygen is used up. When pumping is resumed, the hydrogen sulfide is flushed out (see Figures 3-6 and 3-7).

Routine testing for residual dissolved oxygen (DO) is a simple and useful step to monitor for SRB problems. A dramatic decline in DO may be taken as an early

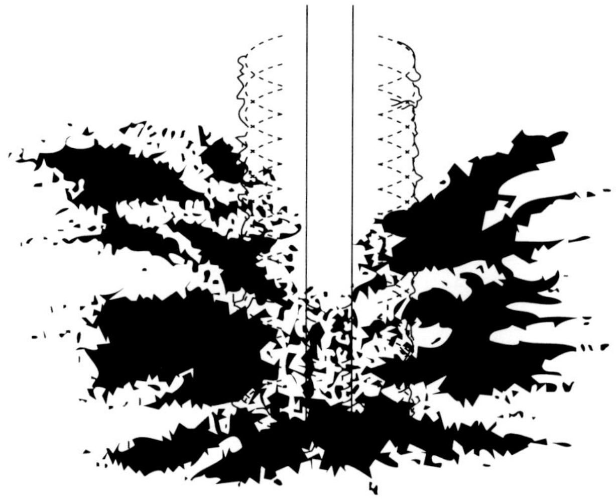

NOTE: When oxygen is used up, hydrogen sulfide spreads through to the well.
This is flushed out when the pumping starts again.

Source: D.R. Cullimore.

Figure 3-7 Operational water well during period of quiescence (not pumping), causing the SRB activity to increase and the hydrogen sulfide to spread (theoretical)

warning signal that conditions are becoming more supportive for SRB growth. Simple analytical methods can be used to monitor the presence and aggressivity of SRB in water.

REFERENCES

American Public Health Association, American Water Works Association, Water Pollution Control Federation. 1992. *Standard Methods for the Examination of Water and Wastewater.* Latest edition. Washington, D.C.: American Public Health Association.

Cullimore, D.R. 1990. An Evaluation of the Risk of Microbial Clogging and Corrosion. In *Water Wells Monitoring, Maintenance, Rehabilitation*. P. Howsam, ed. pp 25–34. London: E. & F.N. Spon.

Pfennig, N., F. Widdel, and H.G. Truper. 1991. The Dissimilatory Sulfate Reducing Bacteria. In *The Procaryotes*, Vol. 1. pp. 926–940. New York: Springer-Verlag.

Postgate, J.R. 1984. *The Sulphate Reducing Bacteria*. 2nd ed. Cambridge, England: Cambridge University Press.

Tiller, A.K. 1990. Biocorrosion in Civil Engineering. In *Microbiology in Civil Engineering*. P. Howsam, ed. pp. 24–38. FEMS Symposium No. 59. London: E. & F.N. Spon.

Widdel, F. and N. Pfennig. 1984. Dissimilatory Sulfate or Sulfur-Reducing Bacteria. In *Bergey's Manual of Systematic Bacteriology.* Vol. 1. pp. 663–679. N.R. Krieg and J.G. Holt, eds. Baltimore, Md.: Williams & Wilkins.

Chapter **4**

Nitrifying Bacteria

Nitrifying bacteria are a group of microorganisms capable of converting ammonia and other reduced nitrogen compounds to more oxidized nitrogen compounds, e.g., nitrate. This process is referred to as nitrification. There are two major groups of nitrifiers (heterotrophs and chemolithotrophs), which are differentiated based on how they obtain energy for growth. This chapter discusses only the chemolithotrophs because of the water quality problems they cause in distribution systems.

Nitrifying bacteria can be helpful in biologically active filtration by decreasing unwanted ammonia entering a treatment plant. However, these bacteria can become a nuisance in some distribution systems where chloramines are used for disinfection. Partial or incomplete nitrification may produce excess nitrite, accelerating the breakdown of chloramines. Chlorine and ammonia may decrease to the extent that disinfection becomes ineffective and regrowth occurs.

BIOLOGY AND ECOLOGY

Nitrifying bacteria are found in a variety of shapes, including rods, curved rods, spheres, spirals, and lobular forms. See Figure 4-1 for an electron micrograph of a nitrifying bacterium. These bacteria range in size from approximately 0.3 to 1.7 μm.

Chemolithotropic nitrifying bacteria are gram-negative and strictly aerobic. Chemolithotropic means that these organisms can obtain energy from oxidizing simple inorganic compounds, such as ammonia or nitrite, and use dissolved carbon dioxide to make sugars for cell material. Many other bacteria commonly found in water obtain their energy from organic compounds. Those organisms are known as heterotrophic bacteria.

Within the cells of nitrifying bacteria are membranes that are believed to be where the nitrification process occurs (Watson, Valos, and Waterbury 1981). In order to successfully complete the nitrification process, two subgroups of nitrifiers (ammonia oxidiziers and nitrite oxidizers) are necessary. Ammonia-oxidizing bacteria (AOB) convert ammonia to nitrite, and nitrite-oxidizing bacteria (NOB) convert nitrite to nitrate. Because the nitrification process by which they obtain their energy is relatively inefficient, the growth rate of nitrifiers is much slower than that of heterotrophic bacteria (Wood 1986).

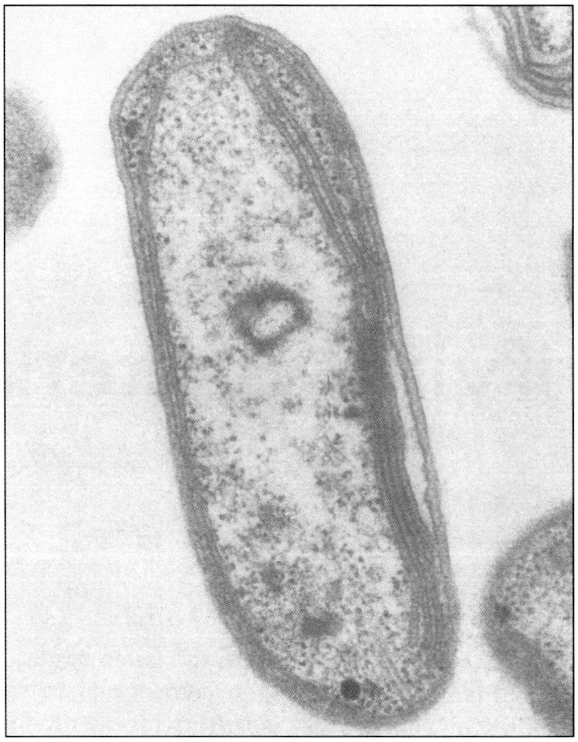

Source: Lieu, Wolfe, and Means (1993).

Figure 4-1 Electron micrograph showing cross section of a nitrifying bacterium

The most common AOB genus recovered from water is *Nitrosomonas* (Painter 1970; Rittmann and Snoeyink 1984; Wolfe et al. 1988 and 1990). Other genera found in water and other habitats include *Nitrosovibrio*, *Nitrosococcus*, and *Nitrospira*. The most common NOB is *Nitrobacter*. Other NOB genera are *Nitrococcus* and *Nitrospina*, which occur in marine habitats.

Nitrifying bacteria can occur in a variety of habitats, including freshwater, marine, soil, brackish water, wastewater, and treated drinking water. Although nitrifiers can survive in a wide range of temperatures (5 to >30°C), their growth rate is controlled by the concentration of ammonia or nitrite, pH, dissolved oxygen, light, and temperature. The optimum pH for the growth rate of nitrifiers is from 7.5 to 8.2 and the optimum temperature is from 25 to 30°C (Bock et al. 1986; Watson, Valos, and Waterbury 1981). Because nitrifiers are sensitive to ultraviolet irradiation, they typically grow in dark environments.

Nitrification is part of a larger process, called the nitrogen cycle, in which elemental nitrogen may be converted to more complex forms of nitrogenous compounds. To complete the loop, these compounds may be converted back to the elemental form of nitrogen gas.

Although there are other groups of organisms and physical means by which nitrification can occur, nitrifying bacteria make up the major group of organisms that can perform nitrification. In addition to the nitrification process, AOB and NOB may play a role in the carbon cycle by excreting organic by-products, which, in turn, support the growth of heterotrophic bacteria (Jones and Hood 1980). Consequently, nitrifiers play a major role in the ecosystem for the existence of life.

Identification

It can be tedious and difficult to identify nitrifying bacteria. This is due, in part, to the slow growth rate of nitrifiers and their poor isolation and colony development on agar culture plates. Photographs of nitrifiers *Nitrosomonas* and *Nitrobacter* are included in the color section of this manual.

Nitrifying bacteria may be identified by observation of growth on selective inorganic media, detection of nitrite or nitrate production, and observation of the physical structure of internal membranes with an electron microscope. A list of media for culturing ammonia-oxidizing and nitrite-oxidizing bacteria is described by Watson, Valos, and Waterbury (1981).

SIGNIFICANCE FOR WATER SUPPLIES

Depending on the treatment application and type of disinfectant, nitrification can be either a benefit or a nuisance to the water industry. Nitrification is beneficial whenever the elimination of natural sources of ammonia from source water is desirable or planned, as ammonia has a considerable chlorine demand.

In the planned approach, nitrifying bacteria are allowed to increase to high levels, typically in the filter beds or the clarifiers, where they convert ammonia to nitrate (Rittman and Snoeyink 1984; Kurtz-Crooks et al. 1986).

Because nitrate does not have an appreciable chlorine demand, complete nitrification results in a decrease in the cost of chlorination. In addition, elimination of free ammonia may lead to increased biological stability of water in the distribution system.

However, partial or incomplete nitrification in distribution systems using chloramines can significantly impact the water quality. As Figure 4-2 indicates, these effects include decay of the chloramine and ammonia residuals and increases in nitrite and heterotrophic bacteria (Wolfe et al. 1988).

Water quality deteriorates because of a buildup of nitrite from the first step of nitrification. The second step, conversion of nitrite to nitrate, either does not occur or occurs at a very slow rate in drinking water. AOBs, which are highly resistant to chloramines, have tolerated monochloramine levels up to 5.0 mg/L (Kirmeyer, Foust, and LeChevallier 1991; Cunliffe 1991; Wolfe et al. 1988 and 1990; Lieu et al. 1993), converting the excess ammonia used to form chloramines into nitrite. NOBs convert nitrite to nitrate at a slower rate and may be more sensitive to antimicrobial agents, including drinking water disinfectants, than AOB (Bock, Koops, and Harms 1986). This leads to an accumulation of nitrite in the water.

Nitrite has a significant chlorine demand, consuming five parts of chlorine for each part of nitrite. In addition, nitrite appears to accelerate the breakdown of the chloramine residual. Decreased disinfectant levels may cause a substantial increase in heterotrophic bacteria, possibly even coliforms. Also, heterotrophic bacteria may increase because they can use the organics secreted by AOB. Thus, nitrification can impair a utility's ability to meet the provisions of the US Environmental Protection Agency's (USEPA's) Surface Water Treatment Rule, which requires utilities to maintain a disinfectant residual or have a heterotropic plate count below 500 colonies/mL in 95 percent of the distribution system samples analyzed per month.

USEPA has also established a maximum contaminant level (MCL) for nitrite at 1.0 mg/L (as nitrogen). Depending on the level of ammonia in the drinking water, the nitrite standard can be exceeded by severe cases of nitrification.

Incomplete nitrification in chloraminated waters occurs predominantly in water with low turnover rates at temperatures greater than 15°C. Typical sites for nitrification problems are in storage tanks and dead ends within the distribution system.

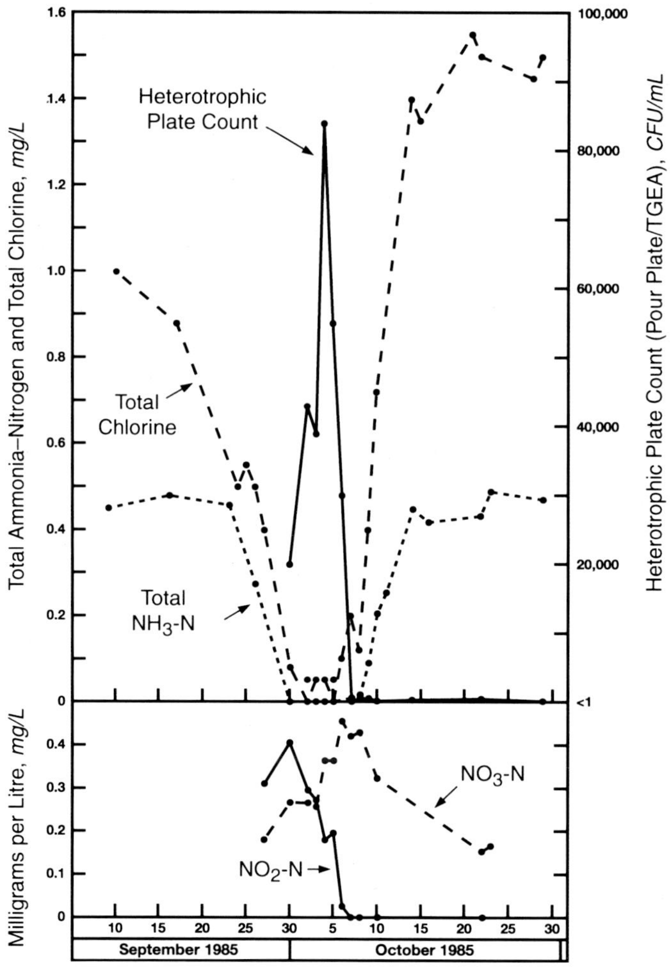

Source: Wolf, Means, Davis, and Barrett (1988).

Figure 4-2 Water quality parameters during a 1985 nitrification episode at Garvey Reservoir, Los Angeles County, Calif.

CONTROL STRATEGIES

Several methods have been used successfully to control nitrification problems in distribution systems containing chloraminated waters. These include flushing the distribution pipeline, superchlorinating reservoirs and storage tanks, increasing the chlorine-to-nitrogen (Cl_2:N) weight ratio, decreasing the detention time in reservoirs and distribution system pipelines (Lieu, Wolfe, and Means 1993), and switching to chlorine disinfection in the distribution system.

Flushing the distribution system is the easiest and quickest method of controlling nitrification. However, it is generally the least effective, particularly when nitrification is severe. Superchlorination will eliminate nitrification because nitrifiers are highly sensitive to free chlorine. This method may be costly and result in the temporary production of elevated disinfection by-products, such as trihalomethanes. Increasing the Cl_2:N weight ratio may prevent nitrification by reducing the amount of excess ammonia available to the AOB (Wolfe et al. 1988; Lieu, Wolfe, and Means 1993). Decreasing the detention time in reservoirs or storage tanks will help prevent increases in the numbers of slow-growing AOB. Finally, periodic switches to free chlorine in the distribution system should prevent biofilm formation that harbors nitrifiers.

The best preventive program involves a comprehensive monitoring system and knowledge of the flow pattern of the water within the distribution system. Parameters that need to be monitored regularly include total and free chlorine, total and free ammonia, nitrite, temperature, dissolved oxygen, and pH (Wolfe et al. 1990). When the system is monitored regularly to detect nitrification early, a greater number of control options are available. In addition, monitoring information helps to optimize chloramine treatment in order to minimize nitrification problems (Lieu, Wolfe, and Means 1993).

REFERENCES

Bock, E., H.P. Koops, and H. Harms. 1986. Cell Biology of Nitrifying Bacteria. *Nitrification*. J.I. Prosser, ed. Washington, D.C.: IRL Press.

Cunliffe, D.A. 1991. Bacterial Nitrification in Chloraminated Water Supplies. *Applied and Environmental Microbiology,* 57:3399.

Jones, R.D., and M.A. Hood. 1980. Interaction Between an Ammonium Oxidizer, *Nitrosomonas* sp. and Two Heterotrophic Bacteria, *Nocardia atlantica* and *Pseudomonas* sp., A Note. *Microbial Ecology,* 6:271.

Kirmeyer, G., G. Foust, and M. LeChavallier. 1991. Optimization of Chloramination for Distribution System Water Quality Control. In *Proceedings of the 1991 Annual Conference in Philadelphia, Pa.* Denver, Colo.: American Water Works Association.

Kurtz-Crooks, J., V.L. Snoeyink, M.D. Curry, and M.L. Reynolds. 1986. Technical Note: Biological Removal of Ammonia at Roxana, Illinois. *Jour. AWWA*, 78:5:94.

Lieu, N.I., R.L. Wolfe, and E.G. Means III. 1993. Optimizing Chloramine Disinfection for the Control of Nitrification. *Jour. AWWA*, 85:2:84.

Painter, H.A. 1970. A Review of Literature on Inorganic Nitrogen Metabolism in Microorganisms. *Water Res.*, 4:393.

Rittmann, B.E., and V.L. Snoeyink. 1984. Achieving Biologically Stable Drinking Water. *Jour. AWWA*, 76:10:106.

Watson, S.W., F.W. Valos, and J.B. Waterbury. 1981. The Family Nitrobacteraceae. *The Prokaryotes*. M.P. Starr et al., eds. Berlin: Springer-Verlag.

Wolfe, R.L., E.G. Means, M.K. Davis, and S.E. Barrett. 1988. Biological Nitrification in Covered Reservoirs Containing Chloraminated Water. *Jour. AWWA*, 80:9:109.

Wolfe, R.L., N.I. Lieu, G.I. Izaguirre, and E.G. Means. 1990. Ammonia-Oxidizing Bacteria in a Chloraminated Distribution System: Seasonal Occurrence, Distribution, and Disinfection Resistance. *Applied and Environmental Microbiology,* 56:451.

Wood, P.M. 1986. Nitrification as a Bacterial Energy Source. *Nitrification*. J.I. Prosser, ed. Washington, D.C.: IRL Press.

This page intentionally blank.

Chapter 5

Nematodes

Nematodes are unsegmented roundworms belonging to the phylum Nematoda, which includes both parasitic and free-living forms. These roundworms, a name that aptly describes their anatomy, are prevalent throughout the environment. It has been said that if the earth and everything on it except nematodes were to disappear, the earth's surface would be outlined by the worms. The top layer of soil can contain over 1 million nematodes per square meter. The greatest number of nematodes are the free-living forms that are commonly associated with soils and aquatic environments.

BIOLOGY AND ECOLOGY

The free-living nematodes found in fresh water are generally microscopic in size, ranging from 5 to 50 μm in width and 100 to 1,000 μm in length. The distinctive characteristics of these animals are their smooth (no cilia) cylindrical shape and whiplike movements that form C- and S-shape outlines. The thrashing motion of these organisms is the result of muscles that run lengthwise and only allow for a side-to-side motion, as shown in Figure 5-1. This method of locomotion is rather ineffective in water but effective in denser materials, such as soils.

The body of a nematode is plump and tapers to a rather broad head and to a tail that may be elongated. The head consists of a lipped mouth, which is followed by an esophagus and intestine that runs the length of the body. The tail is posterior to the anus. Because most nematodes are light beige to clear in color and translucent, they are interesting specimens to view under the microscope. See Figure 5-2.

Nematodes feed on a wide variety of food, depending on the species, including both live and dead plants, animals, and bacteria. Mouth parts of many of the nematodes are specialized for feeding on their choice food. Many of the predatory worms have a rigid, hollow stylet that can extend from the mouth to penetrate the prey and suction cell fluids.

The life cycle of the nematode consists of an egg, several larval stages, and the adult. Larval stages appear similar to the adults except they lack reproductive organs and are smaller. Both male and female are known for most species of nematodes. However, in some cases, reproduction is by the development of an unfertilized egg (parthenogenetic).

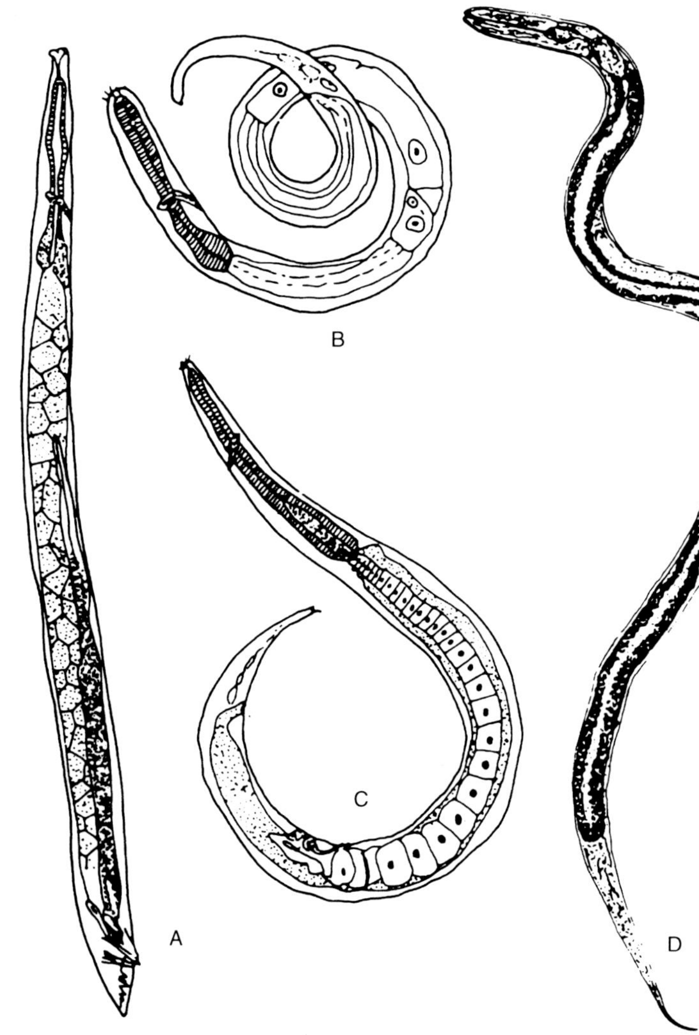

A—Rhabditis (male) (1.6–1.9 mm)
B—Ahcromadora (female) (0.3–3.7 mm)
C—Monhystera (female) (0.8–1.0 mm)
D—Diplogasteroides (female) (1.5–1.85 mm)

Source: Standard Methods for Examination of Water and Wastewater.

Figure 5-1 Roundworms (phylum Nematoda)

Adult females are slightly larger than males. All stages of nematode life, but especially the egg, are resistant to adverse environmental conditions.

Nematodes occur in almost every ecological habitat. In aquatic environments, the greatest numbers are usually found in the aerobic benthic sediments of lakes and rivers. Large numbers are found in wastewater treatment plants and secondary sewage effluent.

The majority of nematodes found in source waters for drinking water plants result from soil runoff, high river flows that suspend the benthic organisms, or from sewage effluent. Nematode levels in source waters generally increase with increasing turbidities and river flow. The distance from a sewage treatment plant also greatly affects the number of nematodes found in the water column, as nematodes flourish in aerobic biological wastewater treatment plants. The number of nematodes in rivers varies

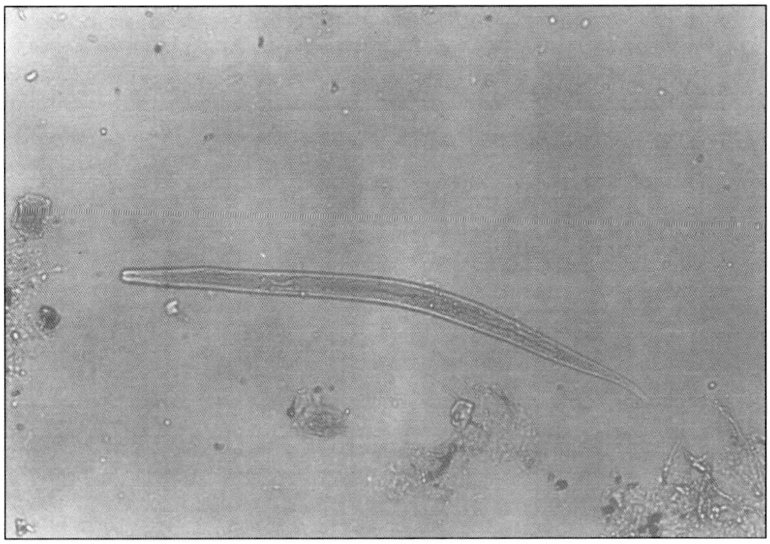

Source: Scott Tighe, Analytical Services Inc.

Figure 5-2 Nematode in bright-field, 200×

greatly depending on the watershed and local sources. A 1979 study by the US Environmental Protection Agency (USEPA) indicated that a direct relationship exists between nematode density in finished water and rainfall, as the scouring action of increased streamflow washes organisms into streams and lakes. River water often contains 0–3 nematodes/L, but on occasion, can reach 800 or more/L.

Identification

Large volumes of water should be examined when sampling for nematodes. A common method of enumeration is to filter 1–20 L of tap water through a 5-μm or smaller membrane filter. A standard coliform bacteria filter of 0.45 μm can be used but will limit the quantity of water examined due to quicker clogging. Place the filter on a moist, absorbent pad, and observe for nematodes and other organisms using a stereoscopic microscope at its highest power (40–50×).

An alternative technique is to rinse the membrane filter with several millilitres of distilled water and examine in a counting dish or Sedgwick–Rafter cell. Live nematodes can be distinguished by observing them for up to a minute to detect movement. A compound microscope is required to identify and count the smallest nematodes (see Figure 5-3). A technique for counting and identifying nematodes may be found in *Standard Methods for the Examination of Water and Wastewater*, Section 10750.

In natural waters, nematodes are often irregularly dispersed and the quantity of organisms can vary greatly over time. For these reasons, it is important to sample large volumes of water over several time periods to adequately assess representative populations in source waters. For turbid raw waters, smaller volumes must be filtered through several membrane filters. Larger pore-size soil sieves (number 500 mesh) or plankton nets (10–35 μm) can be used to concentrate raw waters. Although this will only be a qualitative method, because many of the smaller nematodes will be missed, this method still provides meaningful information. To determine relative numbers, a 35-μm plankton net can be used to collect nematodes from fire hydrants or natural waters.

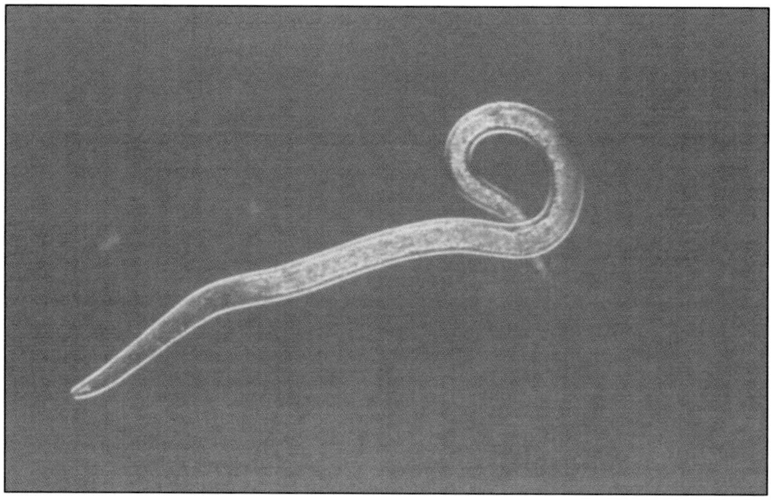

Source: R.C. Lorenz.

Figure 5-3 Nematode at 250×

SIGNIFICANCE FOR WATER SUPPLIES

Nematodes can be found in many drinking waters that have surface water sources. Their density in finished water leaving several water plants in the eastern United States that have monitored for nematodes averaged between 0.13 and 0.4 nematodes/L. Many of the nematodes observed in tap water are alive, as determined by movement. Live nematodes have also been collected from residential faucets and fire hydrants in distribution systems.

The highest density of nematodes in the distribution system is commonly observed when flushing hydrants in low-water-usage, dead-end areas. The greatest numbers of nematodes are frequently noted after generating a scouring velocity by flushing the hydrant. Nematode concentration in these areas are often much greater than in the water leaving the plant.

In many cases, live nematodes are collected in water having a free-chlorine residual. Other aquatic animals are also frequently observed in these samples. In some areas, the nematodes and associated community of organisms appear to be surviving and reproducing in the distribution system.

Nematodes found in source and potable waters have not been identified as pathogenic forms, and they themselves do not present a health threat. However, the presence of live worms does not portray a quality product to the public and may also compromise the microbiological integrity of the drinking water.

Nematodes present in finished water are capable of ingesting human pathogens, such as *Salmonella*, *Shigella*, and viruses. Pathogens ingested by the worm can survive within the worm for several days, which may allow passage into the distribution system. In some areas, the waterborne nematodes may include plant parasitic forms that potentially could have a negative effect when finished water is used for irrigation.

CONTROL STRATEGIES

Conventional water treatment processes are not highly effective in removing nematodes from the supply. Nematodes are very resistant to inactivation by free chlorine and, unless inactivated, nematodes can pass through rapid sand filters. Optimizing the processes of coagulation, settling, lime softening, and chlorination can reduce the numbers of nematodes passing through the plant.

The settling process is the most effective method of reducing nematode concentration in surface water plants. Reservoirs or presettling basins reduce nematodes' numbers from the incoming river water. Within the water plant, the removal of nematodes is greatly aided if the worms can be inactivated prior to reaching the settling basins. Due to the elimination of prechlorination at most water plants because of trihalomethane (THM) concerns, many of the nematodes remain active enough to resist settling. Unchlorinated basins and flumes also can harbor populations of nematodes, algae, and other organisms on the walls and in bottom sediment.

Rapid sand filters do a relatively good job of removing inactivated nematodes, but motile worms are capable of penetrating the filter media. For efficient removal, nematodes should be inactivated prior to filtration. Effective filter backwashing, including air scour and surface wash, will remove inactivated nematodes. Live nematodes found within rapid sand filters generally are at levels equivalent to the applied water, which indicates the worms are not breeding or concentrating in the filters.

The best way to inactivate nematodes is with a strong disinfectant. Unfortunately, nematodes are very resistant to free chlorine, the disinfectant most commonly used in water treatment. Nematodes are three to four orders of magnitude more resistant to free chlorine than bacteria and have a greater resistance than *Giardia* cysts. Consequently, the disinfectant contact times specified in the Surface Water Treatment Rule for *Giardia* inactivation will not effectively remove nematodes. The inactivation of 90 percent of the nematodes in a typical water plant situation with 1 mg/L free chlorine, a pH of 7.8, and a temperature of 10°C requires a contact time of 20 hours or more.

A standard has not been set for nematodes in finished or source water. *Standard Methods for the Examination of Water and Wastewater* states that a 90 percent removal for a water plant using prechlorination, sedimentation, and filtration indicates high treatment efficiency. Further information, references, and a key to freshwater nematodes can be found in *Standard Methods for the Examination of Water and Wastewater*, Section 10750.

REFERENCES

American Public Health Association, American Water Works Association, Water Environment Federation. 1992. *Standard Methods for the Examination of Water and Wastewater.* Latest edition. Washington, D.C.: American Public Health Association.

Chang, S.L. 1961. Viruses, Amoebas and Nematodes and Public Water Supplies. *Jour. AWWA,* 53:288.

Chaudhuri, N., R. Siddiqi, and R.S. Engelbrecht. 1964. Source and Persistence of Nematodes in Surface Waters. *Jour. AWWA,* 56:73.

Lorenz, R.C., N.R. McGill, and T.A. Beer. 1986. The Effects of Free Chlorine on Nematode Mortality. In *Proceedings of the AWWA Water Quality Technology Conference, Portland, Ore.* p. 787. Denver, Colo.: American Water Works Association.

Mott, J.B. et al. 1981. A 13-Month Survey of Nematodes at Three Water Treatment Plants in Southern Ontario, Canada. *Water Res.*, 15:729.

Philpott, L. 1990. Nematodes Discovered in Private Distribution System: Are Biological Standards Appropriate? In *Proceedings of the AWWA Water Quality Technology Conference*; *Advances in Water Analysis and Treatment,* held in Philadelphia, Pa. Denver, Colo.: American Water Works Association.

Tombes, A.S., and A.R. Abernathy. 1979. *Determination of Breeding Sites of Nematodes in a Municipal Drinking Water Facility.* Health Effects Research Laboratory, Office of Research and Development. USEPA. EPA-600/1-79-029.

This page intentionally blank.

Chapter 6

Bloodworms or Midges (Chironomid Larvae)

"Midge fly" and "blind mosquito" are names often given to the adult mosquito-like insect of the species belonging to the two-winged fly Chironomidae (nonbiting midges). Their larvae, which are often referred to as "bloodworms," have infested water distribution systems throughout the world. Due to the great diversity of species, estimated to be from 10,000 to 15,000 worldwide, positive identification of the insects should be left to chironomid specialists.

The information in this chapter is based on conversations with biologists, entomologists, and public health officials, as well as from studies and research papers. It is important to note that successful control of midges depends on determining whether the species is parthenogenetic (reproduces by means of an unfertilized egg) or nonparthenogenetic, and devising controls accordingly.

BIOLOGY AND ECOLOGY

The midge adult usually lives only a few days, although some species survive several weeks. Nonparthenogenetic adult flies mate in aerial swarms or on solid surfaces. Reservoirs, ponds, and streams may contain as many as 100 species, often with an accompanying high density of individuals. Surveys of larvae have revealed larval populations of $50{,}000/m^2$ in certain lakes.

While midges may become a problem in the water treatment plant and distribution system, their presence and condition in surface water are indicators of an ecosystem's health. In unpolluted trout streams, aquatic insects usually comprise 50 to 90 percent of the macroinvertebrate species. Where attempts are made to reclaim lakes, the presence and variety of chironomids are positive indicators of lake health (Fleming 1988).

Because chironomids indicate an ecosystem's health, the organisms are useful indicators of toxic substances in water. After the Sandoz accident, when insecticides were discharged into the Rhine River at Basel, Switzerland, chironomid mortality was observed as far as 500 mi (800 km) downstream (van Urk and Kerkum 1987). Because

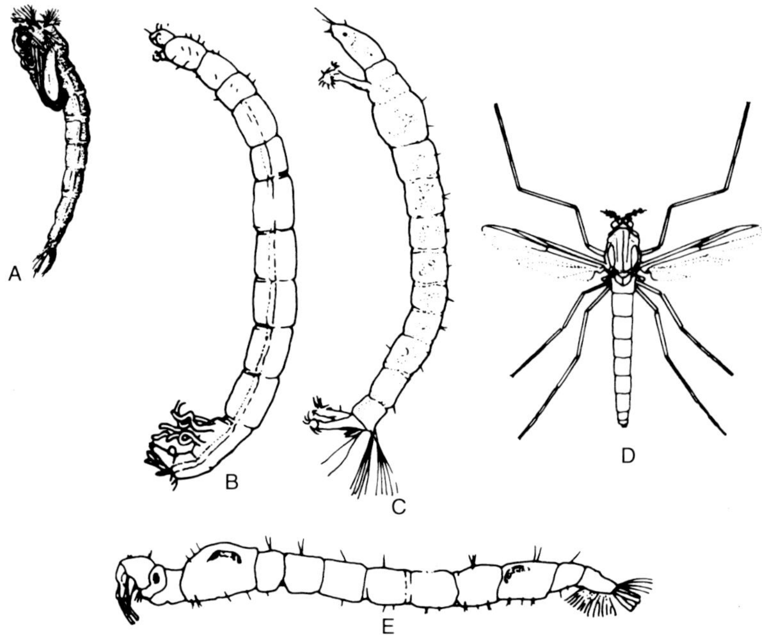

A—Pupa midge, *Chironomus*, Chironomidae
B—Larva midge, *Chironomus*, Chironomidae (5–30 mm)
C—Larva midge, *Ablabesmyia*, Chironomidae (5–10 mm)
D—Adult midge, Chironomidae (4–12 mm)
E—Larva, phantom midge, *Chaoborus*, Culicidae (8–12 mm)

Source: Standard Methods for Examination of Water and Wastewater.

Figure 6-1 Two-winged flies (order Diptera)

midges are sensitive to pollutants and are easily maintained under laboratory conditions, procedures have been developed using chironomids to determine acceptable levels of pollutants. See Toxicity Tests Using *Chironomus*, Sections 8010 and 8750, *Standard Methods for the Examination of Water and Wastewater.*

Identification

The adult chironomid may be as small as 0.5 mm, depending on the species, and resembles a mosquito. The life cycle consists of four stages—egg, larva, pupa, and adult—illustrated in Figure 6-1. Photographs of chironomid may also be found in the color section of this manual.

The adult chironomid lays its eggs in a gelatinous mass containing from 3 to 3,000 eggs, depending on the species. Each mass of eggs may be laid on the surface of the water, attached to aquatic plants or stones at the water's edge, or attached to the sides of reservoirs, aerators, basins, or elevated storage facilities. Egg masses that are not attached to objects will sink to the bottom. On hatching, the larvae disperse by undulating in a rapid motion.

After dispersion, the larvae construct silken tubes around themselves, shown in Figure 6-2, with an open end that enables them to feed on nutrients in the water. The food supply may be decaying vegetable matter, algae, protozoans, live or dead bacteria, and detritus. Larvae may be blood red, green, yellow, brown, or transparent. Their size varies from less than 1 mm to slightly more than 2.5 mm. The larval stage may last from weeks to several years, depending on food supply, temperature, and species.

Source: Drs. J.H.M. van Lieverloo, Kiwa N.V. Research and Consultancy, Nieuwegein, The Netherlands.

Figure 6-2 Pupa of a chironomid (Insecta, Chironomidae), 20×, dorsal view

Transformation from larvae to pupae takes place in the silken tube. The pupal stage normally lasts approximately three days. Pupae leave the tube and swim to the surface just before the adult emerges.

SIGNIFICANCE FOR WATER SUPPLIES

The chironomid *Paratanytarsus grimmii* was first recognized in the 1930s (Langton, Cranston, and Armitage 1988). At that time, the only known effective removal method occurred when water mains froze. The species was noted to be widely distributed in eastern and southern England. In 1986 an infestation occurred in Cyprus. Since that time, evidence has accumulated indicating that the infestation problem is widespread in other areas of the world.

In most cases these infestations are not recognized as a public nuisance until the chironomid population increases dramatically (Bay 1989). These increases may correlate with cleaning screens, replacing old filters, and flushing systems. Midge larvae may penetrate the filters in the treatment plant and enter the distribution system. Populations may establish themselves and appear at the tap when the population peaks or is disturbed by increased flushing.

A survey sent by Lowell Water Utility, Lowell, Ind., to state health departments in the southern portions of the United States indicates that almost all have experienced midge larvae problems. Excerpts from this correspondence follow, and other documented infestations of midges are listed in references for this section.

- Washington County, N.C. Midge flies apparently penetrated the screened enclosure surrounding an aeration tower. Concentrated efforts to disinfect the tower reservoir and splash boards did not rid the system of larvae. Eventually, filters were constructed and the larvae were eliminated from the distribution system.
- Gaston County, N.C. The intrusion of larvae into the distribution system was prevented by backwashing the filters of the water treatment plant at frequent intervals. The perpetual jerking motion of the larvae had allowed them to penetrate the filter. After increasing filter washing to every eight hours, no larvae were observed passing through the filters.
- Missouri. Midge fly larvae infested two unchlorinated and two chlorinated sections of a city's clearwell. The treatment supervisor was instructed by state health officials to increase the chlorine residual from 2.5 ppm to 4 or 5 ppm with sodium hypochlorite and to make the ventilators insectproof. The operator also installed an electric insect trap to kill any adult flies that might emerge. These strategies were apparently successful.

CONTROL STRATEGIES

The key to controlling nonparthenogenetic midges is to exclude adult midge flies from all potable water treatment areas. If an infestation occurs, the only way to rid the system of this nuisance is to eliminate the swarming and mating behaviors, and ultimately the deposition of eggs.

The control of parthenogenetic midges is more difficult. Controlling the swarming and mating behavior will not prevent infestation in this case. To further complicate this situation, female chironomids may develop and deposit fertile eggs within the pupa, without undergoing emergence (Langton, Cranston, and Armitage 1988) (see Figure 6-3).

Chemical control methods have been most successful with parthenogenetic midges. Although, at the time of this writing, new areas of control were under investigation, their success will not be known for some time. It is the author's opinion that the control of parthenogenetic midges is beyond the resources available to small and medium-sized utilities. Such systems are best served by enlisting the aid of experts.

D.A. Williams (1974) gives an account of the infestation of a parthenogenetic midge capable of reproducing within distribution mains without passing through the adult stage. The infestation occurred in 1971 in Essex, England. The usual methods of cleaning reservoirs, installing electronic insect killers, and making enclosures insectproof were not successful in eliminating midge fly larvae. Essex Water Company even installed microstrainers in an unsuccessful attempt to eliminate the larvae.

In June 1973, it was discovered that the chironomid was parthenogenetic. After meetings with water company officials and health department medical officers, a chemical control using pyrethrins was applied to the affected area. After dosing the system with 0.025 mg/L pyrethrins, an insecticide whose use is limited in the United States, only one complaint was received concerning dead larvae in the tap water.

In 1988 the town of Lowell, Ind., experienced a similar infestation of a midge identified by Dr. Martin B. Berg, University of Notre Dame, to be parthenogenetic. It is believed that the infestation started in an elevated storage tank serving the eastern portion of town. Storage tanks were cleaned and disinfected several times. High concentrations of chlorine had little effect on the larvae. The standpipe portion of the tower was disinfected with over 50 mg/L of chlorine for 48 hours, and yet some larvae

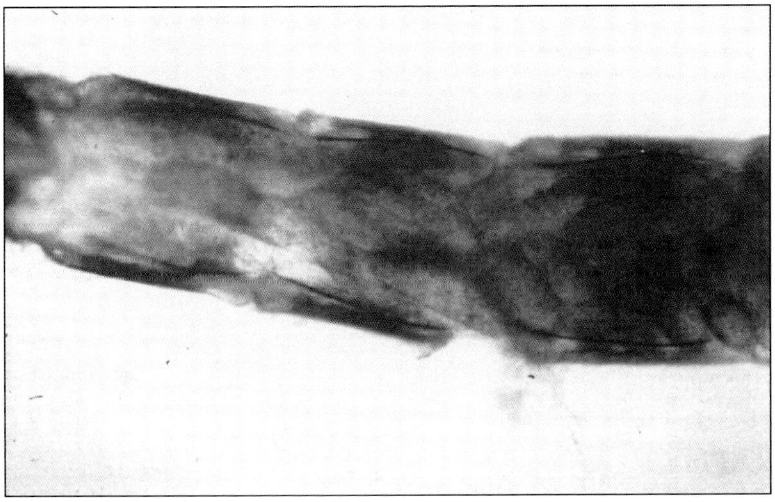

Source: Drs. J.H.M. van Lieverloo, Kiwa N.V. Research and Consultancy, Nieuwegein, The Netherlands.

Figure 6-3 Parthenogenetically formed eggs in abdomen of a chironomid pupa (Insecta, Chironomidae), ventral view, 70×

were still alive when the tower was drained. In short, no effective controls have been found at this writing for parthenogenetic midge infestations of water supplies.

A novel approach is being used at Lowell to determine whether midge larvae can be controlled through various methods of limiting their food supply. Another method of control that is experiencing some success in laboratory tests is the use of cationic polymer. Tests conducted at Michigan State University's Department of Entomology indicate cationic polymers used in water clarification may successfully control parthenogenetic midges. Adding zinc orthophosphate may provide additional controls in the distribution system.

A system experiencing midge problems should contact specialists to identify the organism at the first sign of infestation. Specific identification can be made by only a few specialists in this field. It is imperative that utility personnel know what type of midge they are dealing with in order to formulate an organized plan to rid the system of the insect. As mentioned, if the infestation is from a nonparthenogenetic chironomid, the avenues available to the operator to eliminate the organism are somewhat more favorable.

All treatment plant structures should be made insectproof with fine-mesh screen or furnace filter material. Even 20×20 mesh screen may not be fine enough to exclude some species of chironomid. It may be necessary to incorporate vacuum vents in elevated or ground storage facilities so that plugged vents do not cause a vacuum and collapse the structure. Covering filters and reservoirs with fine screening material and installing electric insect traps may help eliminate the adult midge fly. All structures should be thoroughly cleaned; however, high concentrations of chlorine have had limited success.

REFERENCES

Alexander, M., and R.W. Merritt. Not yet published. Project Report on the Control of *Paratanytarsis grimmii* Infesting the Lowell, Indiana Drinking Water Distribution System. East Lansing, Mich.: Michigan State University Department of Entomology.

Ali, A. Nuisance Chironomids and Their Control: A Review. Agricultural Research and Education Center, *JFAS*.

Ali, A., and P.K. Chaudhuri. 1988. Evaluation of a New Benzoylphenylurea Insect Growth Regulator (US 84572) Against Chironomid Midges in Experimental Ponds. *J. Am. Mosquito Control Assoc.*, 4:542–544.

American Public Health Association, American Water Works Association, Water Environment Federation. 1992. *Standard Methods for the Examination of Water and Wastewater.* Latest edition. Washington, D.C.: American Public Health Association.

Arnold, G.E. 1936. Plankton and Insect Larvae Control in California Waters. *Jour. AWWA,* 28:1469.

Bahlman, C. 1932. Larval Contamination of a Clear-Water Reservoir. *Jour. AWWA,* 24:660.

Bay, E. 1989. Personal communication.

Brown, K.W. 1933. Experiences With Well Water in Uncovered Reservoir. *Jour. AWWA,* 25:337.

Burfield, I., and D.N. Williams. 1975. Control of Parthenogenetic Chironomids With Pyrethrins. *Water Treat. Exam.,* 24:57–67.

Coffman, W.P., and L.C. Ferrington Jr. *Chironomidae.* Univ. of Pittsburgh and State Biological Survey of Kansas.

Fleming, C.R. 1988. Effects of Reclamation on an Urban Winter-Kill Lake at Winona, Minnesota. Internationale Vereinigung fuer Theoretische und Angewandte Limnologie Verhandlungen *IVTLAP,* 23:1:341–345.

Flentje, M.E. 1945. Control and Elimination of Pest Infestation in Public Water Supplies. *Jour. AWWA,* 37:1194.

Flynn, T., and P.M. Bolas. 1985. A Simple Method of Chironomid Control at Water Treatment Works. *Inst. Water Eng.,* 39:5:414–422.

Hechmer, C.A. 1932. *Chironomus* in Water Supply. *Jour. AWWA,* 24:665.

Ingram, W.M., and A.F. Bartsch. 1960. Animals Associated With Potable Water Supply. *Jour. AWWA,* 52:1521–1550.

Langton, P.H. 1974. On the Biology of the Parthenogenetic *Paratanytarsus* Breeding in the Essex Water Company Distribution System. *Water Treat. Exam.,* 23:230–231.

Langton, P.H., P.S. Cranston, and P. Armitage. 1988. The Parthenogenetic Midge of Water Supply Systems *Paratanytarsus grimmii* Schneider (Diptera: Chironomidae). *Gull. Entomol. Res.,* 78:317–328.

Mackenthun, K.M. 1973. *Toward a Cleaner Aquatic Environment.* Washington, D.C.: US Environmental Protection Agency.

Mundie, J.H. 1956. The Biology of Flies Associated With Water Supply. *J. Inst. Water Eng.,* 55:178–190.

Mundie, J.H. 1957. The Ecology of Chironomidae in Storage Reservoirs. Trans. *Roy. Entomol. Soc.* London. 109-149-232.

Silvey, J.K.G. 1956. Bloodworms in Distribution Systems. *Jour. AWWA,* 48:275–280.

van Urk, G., and F.C.M. Kerkum. 1987. Chironomid Mortality After the Sandoz Accident and Deformities in *Chironomus* Larvae Due to Sediment Pollution in the Rhine. *Aqua.* International Water Supply Association. Rijkswaterstaat, Institute for Inland Water Management and Waste Water Treatment. Pergamon Press: Lelystad, The Netherlands. 4:191–196.

Walker, E.D., E.J. Olds, and R.W. Merritt. 1988. Gut Analysis of Mosquito Larvae (Diptera: Colicidae) Using DAPI Stain and Epifluorescence Microscopy. *J. Med. Entomol.,* 25:551–54.

Walshe, B.M. 1951. The Feeding Habits of Certain Chironomid Larvae (Subfamily Tendipedinae). *Proc. Zool. Soc.,* London, England. 121:63.

Williams, D.A. 1974. An Infestation by a Parthenogenetic Chironomid. *Water Treat. Exam.,* 23:215–219.

Chapter 7

Crustacea

The class Crustacea contains several aquatic species that have been recognized as nuisance organisms in potable water supplies. Although these organisms are not considered public health risks, they are often the source of consumer complaints. Some of the larger species can be seen with the unaided eye, but even the smaller organisms are often conspicuous because of their active movements. Crustacea have been associated with water discoloration and taste-and-odor complaints.

BIOLOGY AND ECOLOGY

Taxonomic Listing

A taxonomic listing of the various orders of Crustacea encountered in potable water, with representative genera (Levy, Hart, and Cheetham 1986), follows.

- Phylum—Arthropoda
  - Class—Crustacea
    - Order—Cladocera
      - Genus *Bosmina* spp.
      - Genus *Daphnia* spp.
    - Order—Copepoda
      - Genus *Cyclops* spp.
    - Order—Isopoda
      - Genus *Asellus* spp.
    - Order—Amphipoda
      - Genus *Gammarus* spp.
      - Genus *Hyallela* spp.

Figures 7-1 and 7-2 illustrate some of the organisms discussed in this chapter.

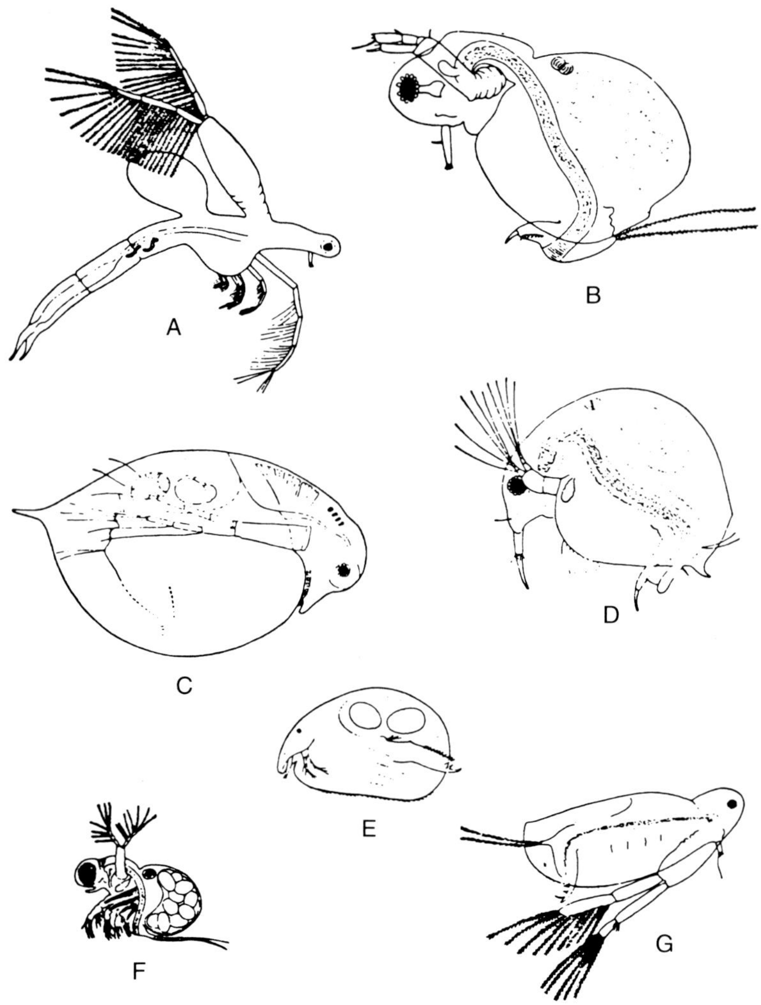

(Phylum Arthropoda, Class Crustacea): Types of cladocerans (Order Cladocera).

A—*Leptodora* (9 mm)
B—*Moina* (1.5 mm)
C—*Daphnia* (2 mm)
D—*Bosmina* (0.4 mm)
E—*Alona* (0.4 mm)
F—*Polyphemus* (1.5 mm)
G—*Diaphanosoma* (1.5 mm)

Source: Standard Methods for Examination of Water and Wastewater.

Figure 7-1 Crustaceans

Cladocera

The order Cladocera, whose members are commonly referred to as water fleas, range in size from 0.25 to 3.0 mm in length. The carapace, a hard shell-like covering, encloses the trunk, but does not extend over the head. The head projects ventrally and somewhat posteriorly, giving the appearance of a bird with a short beak. Members of the representative genus *Daphnia* are often characterized as resembling a true flea, which is an insect, and thus the colloquial name for the group. See the color reference section for a line drawing and photographic comparison of *Daphnia*. Species of the genus *Bosmina*, shown in Figure 7-3, are also found in potable supplies.

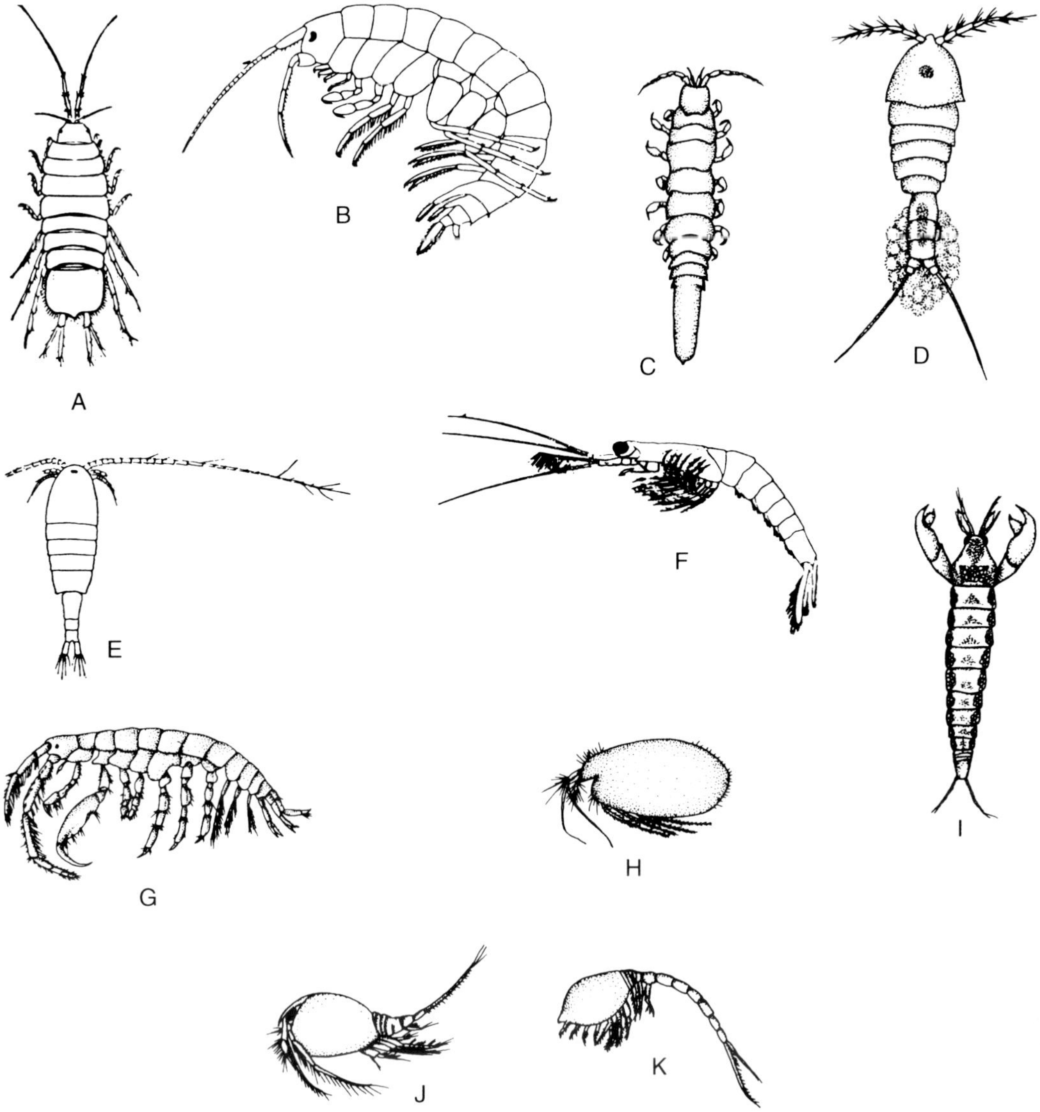

(Phylum Arthropoda, Class Crustacea). Selected crustacean orders.

A—Sowbug, *Asellus*, Isopoda (2 cm)
B—Scud, *Gammarus*, Amphipoda (1.5 cm)
C—Isopod, *Idotea*, Isopoda, marine (1–3 cm)
D—Copepod, *Tigriopus*, Copepoda, marine (2–4 mm)
E—Copepod, *Diaptomus*, Copepoda (2 mm)
F—Mysid shrimp, *Mysis*, Mysidacea (10–20 mm)
G—Amphipod, *Jassa*, Amphipoda, marine (1.5 cm)
H—Ostracod, *Ostracoda*, marine (2–3 mm)
I—Tanaid, *Anatanis*, Tanaidacea, marine (8–10 mm)
J—Nebaliad, *Epinebalia*, Leptostraca, marine (10–20 mm)
K—Cumacean, *Oxyurostylis*, Cumacea, marine (8–12 mm)

Source: Standard Methods for Examination of Water and Wastewater.

Figure 7-2 Crustaceans

When viewed under a stereoscopic microscope, the bodies of Cladocera members appear translucent. Large compound eyes are found on the head and serve as a prominent feature. Locomotion is by means of powerful antennae, with the organisms swimming in a jerky vertical motion. The majority of members of this order are filter feeders, feeding on plankton and decomposed products. Reproduction is normally sexual, but under certain conditions the genus can undergo parthenogenesis, or asexual

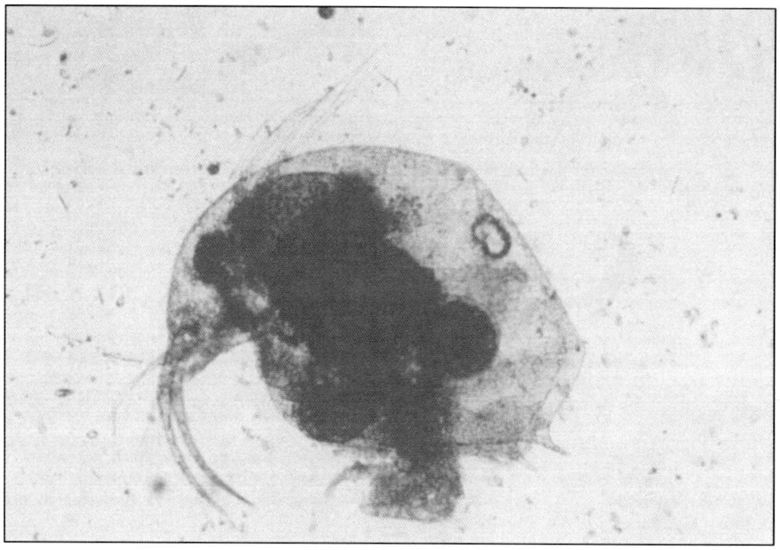

Source: K. Hancock, CHDiagnostic & Consulting Service Inc.

Figure 7-3 Crustacean *Bosmina* with eggs, iodine stain, bright-field microscopy, 60×

reproduction. Population peaks are generally associated with warm water conditions. Organisms enter the treatment train, but usually are unable to adapt to flowing conditions. Species are found alive in service reservoirs, but not in the mains. Apparently the organisms do not multiply in the mains.

Copepoda

The genus *Cyclops* serves as the representative species of the order Copepoda. Members of this group are similar in size range to Cladocera; however, the vast majority are less than 1 mm in length. The body is normally short and cylindrical, consisting of a head, trunk, and abdomen. The thoracic region is normally tapered and has no appendages. The head has two sets of antennae, the first set longer than the second. In females conspicuous egg sacs are found attached laterally to the body. A line drawing and photograph of *Cyclops* appears in the color section of this manual.

Cyclops, as the name suggests, has a single eye in the center of the head region. Reproduction is sexual with breeding peaks occurring in the summer months. Members of the order may be herbivorous or carnivorous. Organisms enter the system through the treatment train and are capable of breeding in the mains. Infestations have been reported even in filtered supplies and have been attributed to eggs passing through the filters and hatching in the mains.

Isopoda

Organisms in the order Isopoda are often referred to as water lice or aquatic sow bugs. The representative genus is *Asellus*, shown in Figure 7-4 and in the color section. The body is dorsoventrally flattened with the head often being shield-shaped. Sizes range from 5 to 15 mm in length with distinct abdominal segments. The organisms are grayish or brownish in color. Members have seven pairs of legs with the posterior appendages being longer than the anterior appendages. The first pair of legs contain a grasping apparatus. Reproduction is sexual with spring and autumn breeding peaks. Population peaks often occur in summer.

Source: Drs. J.H.M. van Lieverloo, Kiwa N.V. Research and Consultancy, Nieuwegein, The Netherlands.

Figure 7-4 *Asellus aquaticus* (Crustacea, Isopoda), dorsal view, 5×

The organisms are capable of breeding in mains. Isopoda live at the bottom of bodies of water, generally move by crawling, and are detritus feeders. Owing to their relatively large size compared with previously mentioned crustacea, Isopoda infestations often elicit consumer complaints.

Amphipoda

Members of the order Amphipoda resemble the Cladocera. However, the body is generally compressed along the side. These organisms are often referred to as fresh water shrimp. Amphipods are similar to Cladocera in size. These organisms reproduce sexually with breeding peaks generally occurring during the spring and population peaks occurring during the summer months. Amphipods are capable of breeding in the mains. These organisms live at the water's bottom, move by crawling, and feed on detritus. *Gammarus* and *Hyallela* are representative genera for this order. See Figure 7-5.

SIGNIFICANCE FOR WATER SUPPLIES

The majority of crustaceans enter a water system through the treatment plant, and some have the capacity to breed in the mains. Larger numbers of crustaceans, both adult and larval forms, can block rapid gravity filters and disrupt filter operation, increasing turbidity as a result of floc breakthrough. Several investigators have suggested that members of the class Crustacea may harbor coliform bacteria and provide a means of protection from the disinfection process.

CONTROL STRATEGIES

As mentioned, crustaceans normally enter the distribution system through the treatment plant. They are most often associated with unfiltered systems supplied from open reservoirs. These occurrences are not limited to unfiltered supplies. The organisms or their eggs have been shown to penetrate filters, especially when filters operate improperly.

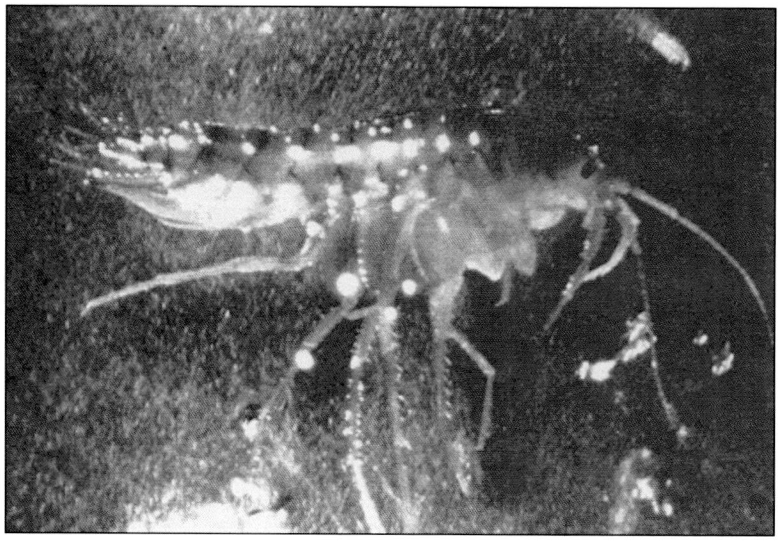

Source: Richard Levy.

Figure 7-5 The amphipod *Hyallela azteca* (1.0 cm) isolated from a point-of-use filter

The organisms tend to concentrate in dead ends and areas of low flow. A combination of factors, including sediment accumulation and dissipation of disinfectant residual, contribute to the proliferation of crustaceans, particularly those that are capable of breeding in the mains. The presence of dead and decaying organisms can result in water quality deterioration.

A systematic flushing program is recommended as both a preventive and active control measure. Filtration of all surface water supplies may greatly limit the problem of macroinvertebrate infestations. Upgrading existing filtration efficiency can improve organism control significantly.

Disinfectants vary in their ability to inactivate these animals. *Cyclops* spp. are reported to be more sensitive to chloramination than to free chlorine, while the opposite is true for *Bosmina* and *Daphnia* species. Isopods and amphipods are often quite resistant to chemical inactivation.

REFERENCES

Barnes, R.D. 1974. *Invertebrate Zoology.* 3rd ed. Philadelphia, Pa.: W.B. Saunders Co.

Levy, R.V., F.L. Hart, and R.D. Cheetham. 1986. Occurrence and Public Health Significance of Invertebrates in Drinking Water Systems. *Jour. AWWA,* 78(9):105–110.

Mackenthun, K.M., and L.E. Keup. 1971. Control of Biological Problems Encountered in Water Supplies. *Jour. AWWA,* 65:520–526.

Mitchan, R.P., M.W. Shelley, and C.M. Wheadon. 1983. Free Chlorine Versus Ammonia-Chlorine: Disinfection, Trihalomethane Formation and Zooplankton Removal. *Jour. AWWA,* 75(4):196–198.

Smalls, I.C., and G.F. Greaves. 1968. A Survey of Animals in Distribution Systems. *J. Water Treat. & Exam.*, 19(3):150–183.

Chapter 8

Rotifers

Rotifer is the common name for a large group of microscopic aquatic invertebrates. Approximately 1,500 species of rotifers have been identified, the majority of which are free living. They usually inhabit the zones between high- and low-level watermarks of ponds, lakes, and other freshwater bodies around the world. A few species are found in salt water.

BIOLOGY AND ECOLOGY

Rotifers are often associated with a bountiful food supply, along with other favorable factors. Their presence is useful in evaluating conditions in streams. Unlike plankton, which do not respond fully to pollution for a considerable distance downstream, the condition of rotifers is determined by the water quality at that point. Rotifers also may be indicative of abundant dissolved oxygen and little to no stratification. They are considered a less expensive toxicity testing measurement than brine shrimp because their eggs can be dried and reconstituted on demand. Their eating habits are highly sensitive to aquatic contaminants, allowing their use as highly effective toxicity yardsticks.

Rotifers are multicellular organisms, from 100 to 600 μm long, which sometimes constitute a considerable percentage of total freshwater plankton. *Brachionus plicotilis* is commonly used as a start food for marine fish larvae. Rotifer populations fluctuate widely, depending on the particular species present, available light, the season, and many other factors that are not yet understood. Populations of certain rotifers wax and wane a number of times a year, while others only become abundant once or twice a year.

Identification

Physical identification of rotifers is made easier by the presence of such unique features as a funnel-like structure at the front of the organism bearing circles of cilia (short, hair-like fringes) that often give the appearance of rapidly revolving wheels (see Figure 8-1). This phenomenon is responsible for the name rotifer or "wheel bearer." Water currents set up by the beating cilia bring minute food particles into the rotifer's mouth, which is at the center of the corona. The cilia also serve a dual function as organs of locomotion.

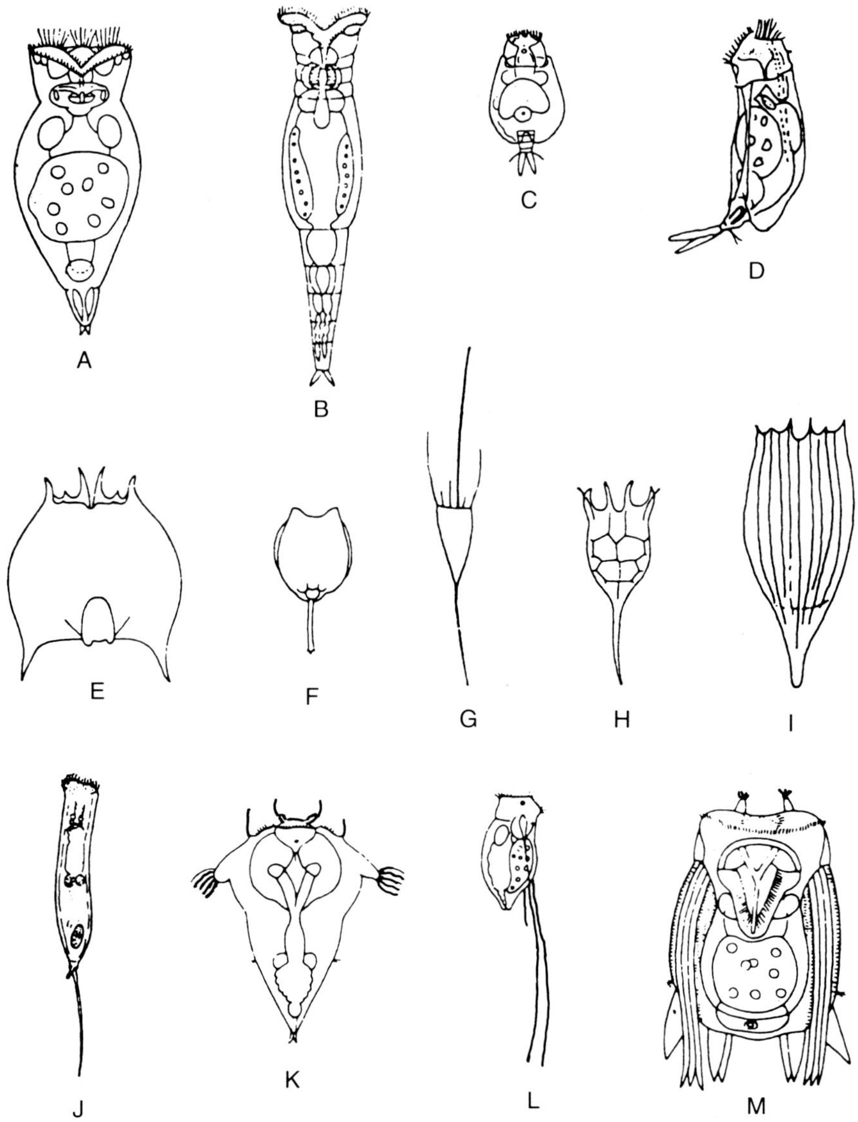

A—*Epiphanes*	(600 µm)	H—*Keratella*	(200 µm)
B—*Philodina*	(400 µm)	I—*Notholca*	(200 µm)
C—*Euchlanis*	(250 µm)	J—*Trichocerca*	(600 µm)
D—*Proales*	(450 µm)	K—*Synchaeta*	(260 µm)
E—*Brachionus*	(200 µm)	L—*Filinia*	(150 µm)
F—*Monostyla*	(150 µm)	M—Polyarthra	(175 µm)
G—*Kellicottia*	(1 mm)		

Source: Standard Methods for Examination of Water and Wastewater.

Figure 8-1 Rotifers (phylum Rotifera)

Rotifers have bilaterally symmetrical cylinder- or vase-shaped bodies the surface of which is covered by a cuticle that, in some genera, is rigid and called a lorica. These "suits of armor" are often found in water as skeletal remains. A characteristic organ in the gut, the mastax seen in Figure 8-2, is a grinding organ that can be easily observed through the body wall. The posterior end of many rotifers is a forked foot with two projections called "toes." The color of the organisms vary according to the contents of the stomach and certain other organs.

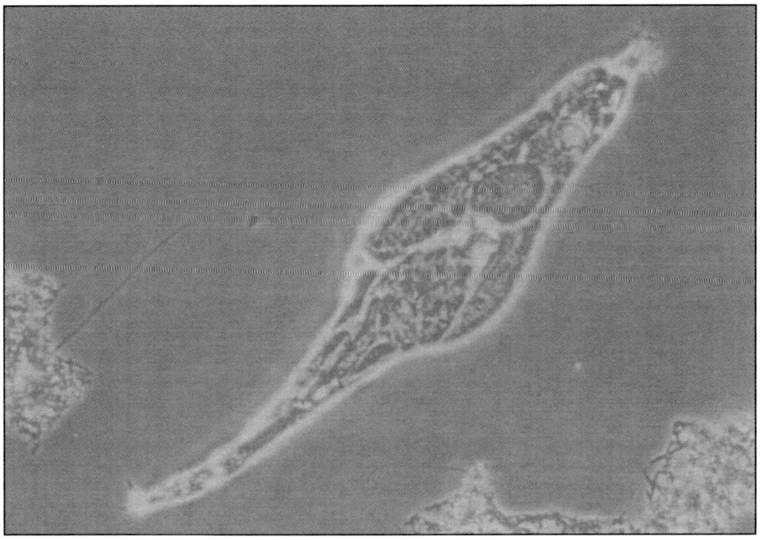

Source: M. Richard, PhD, RBD Inc., Ft. Collins, Colo.

Figure 8-2 Rotifer, 200×

SIGNIFICANCE FOR WATER SUPPLIES

Rotifers have been associated with uncovered finished water reservoirs. They may also be found in the treatment system, on filter surfaces, and in basins. These organisms are capable of detaching from floc particles using their rotary organs, and small, compact rotifers may be transported through filter media.

Rotifers' presence in groundwater, determined through particle analysis, along with other microscopic organisms, such as diatoms, coccidia, plant debris, insect parts, and *Giardia* cysts, serve as indicators of surface water contamination of groundwater systems.

CONTROL STRATEGIES

Although rotifers may occur in extremely large numbers in source waters, water treatment plant personnel may be only vaguely aware of their existence. There are several reasons for this. Despite their widespread distribution, rotifers are not seen under normal circumstances because they are microscopic. They have not been implicated as dangerous pathogens and are viewed more as a laboratory curiosity, or at worst, as an annoying pest.

Another reason rotifers receive so little attention is because treatment need not be directed specifically toward their removal. The destruction or removal of rotifers is incidental to the elimination of the smaller and potentially harmful pathogens that may be present in the water. Optimizing coagulation, flocculation, and filtration remains the "best available technology" for the removal of these organisms.

To remove over 90 percent of rotifers, they should be deactivated before entering the treatment train by adding chlorine (1.5–2 mg/L for 1 min), potassium permanganate (0.5–1 mg/L for 15–20 min), or by using ultrasonic waves for a reaction time of a few seconds. The least desirable of these methods is chlorine, because of the formation of organochlorides. When rotifers are inactive, they are more likely to be removed by flocculation, sedimentation, and filtration.

Chemical pretreatment (using coagulants and polymers) and filter-media pore sizes are designed so that conscientious operation of a well-maintained facility will remove rotifers. Their removal during water treatment is dependent on their shape, size, and mobility. Their presence in treated water most often indicates contamination of the filtered water channels or a defect in the treatment train.

REFERENCES

Alfred, E.M. 1968. *Zoology*. New York.

American Public Health Association, American Water Works Association, Water Pollution Control Federation. 1992. *Standard Methods for the Examination of Water and Wastewater*. Latest edition. Washington, D.C.: American Public Health Association.

American Water Works Association. 1973. AWWA Manual M20, *Water Chlorination Principles and Practices*. Denver, Colo.: American Water Works Association.

Barnes, R.D. 1987. *Invertebrate Zoology*. 5th ed.

Bernhardt, H., and B. Lusse. 1989. Elimination of Zooplankton by Flocculation and Filtration. *Aqua (Journal of Water Supply Research and Technology)*, 38:1:23–31.

Donner, J. 1966. *Rotifers*. New York: Warne.

Engeman, J.G., and R.W. Heyner. 1981. *Invertebrate Zoology*.

Evans, M.S. 1980. *Lake Huron Crustacean and Rotifer Zooplankton*. Michigan University Special Report. East Lansing, Mich.: Michigan State University.

Glowing Microbes May Hold Key to More Effective Toxicity Testing. 1993. *Environment Today*, 4:1:1,23.

Hoffbuhr, J.W., J. Blair, M. Bartleson, and R. Karlin. 1986. Use of Particulate Analysis for Source and Water Treatment Evaluation. In *Proceedings of the AWWA Water Quality Technology Conference: Advances in Water Analysis and Treatment* held in Portland, Ore. Denver, Colo.: American Water Works Association.

Hutchinson, G.E. 1967. *A Treatise on Limnology*.

Ingram, W.M., and A.F. Bartsch. 1960. AWWA Manual M7, *Animals Associated with Potable Water Supplies, Operator's Identification Guide*. Denver, Colo.: American Water Works Association.

Kirk. Rotifer Keratella. *Limnology and Oceanography*, 88:33:1440–50.

Korstad, J. 1987. *Biology of Brachionus picatilis*. Royal Norwegian Council for Scientific and Industrial Research, Norway.

Lusse, B., and J. Clasen. 1991. Studies on the Planktonic Rotifer *Notholca Caudata* with Regard to Drinking Water Purification. *Aqua (Journal of Water Supply Research and Technology)*, 40:6:380–384.

Scarfe. 1981. Population Increases of Rotifers. *Biologist*, AG 81 66:82–91.

Sladecek, V. 1983. Rotifers as Indicators of Water Quality. *Hydrobiologia*, 109:169.

Vasconcelos, J., T. Notestine, J. Hudson, and J. Pluntze. 1990. Use of Particulate Analysis and Other Parameters in the Evaluation of Subsurface Water Sources. In *Proceedings of the AWWA Water Quality Technology Conference: Advances in Water Analysis and Treatment* held in Philadelphia, Pa. Denver, Colo.: American Water Works Association.

Chapter 9

Zebra Mussels

The zebra mussel, *Dreissena polymorpha*, first chronicled by Pallas in 1771, is a freshwater bivalve mollusc, or clam, native to the Caspian Sea region of Eurasia. This mussel was introduced into several European freshwater ports near the end of the 18th century and has since spread, primarily via canals, throughout the inland waterways of Europe.

Rapid transoceanic shipping has provided a new means for species endemic to one continent to move to colonize another. The zebra mussel apparently was introduced to North America when ballast water containing larval zebra mussels from Europe was discharged into the waters of the Great Lakes by a transoceanic freighter sometime during the mid-1980s. Zebra mussels were first discovered in Lake St. Clair in June 1988 and by 1989 had spread prolifically throughout Lake Erie. By fall 1990, they had been found at one or more locations in each of the other four Great Lakes.

BIOLOGY AND ECOLOGY

Floating on water currents for up to two weeks, free-swimming larvae called veligers may be carried great distances before settling onto a hard surface. It is common for more than 95 percent of the veligers to die during this phase of development. Once settled, the surviving veligers attach themselves to a surface by secreting sticky filaments called byssal threads. Over the next few weeks, the attached veligers will take on the typical double-shelled mussel characteristics. Sexual maturity generally occurs early in the second year of life. The typical life span is three to five years.

The reproductive cycle of the zebra mussel, illustrated in Figure 9-1, enables it to spread rapidly and establish large populations. A mature female zebra mussel (2–5 years old) can produce 30,000 to 40,000 eggs per season. Egg production can begin when water temperatures reach 12°C and may continue as long as water temperatures remain above 12°C. The eggs are released into water, where they are fertilized by the males. Several days later, these eggs hatch into veligers.

Because of its ability to securely attach to virtually any hard substrate, coupled with rapid reproductive capability and tolerance for a wide range of environmental conditions, the zebra mussel has become widely disseminated in marine environments. Colony densities exceeding 100,000 zebra mussels/m^2 have been found in western Lake Erie. Veliger larvae of Asiatic clams are far less common in fresh water.

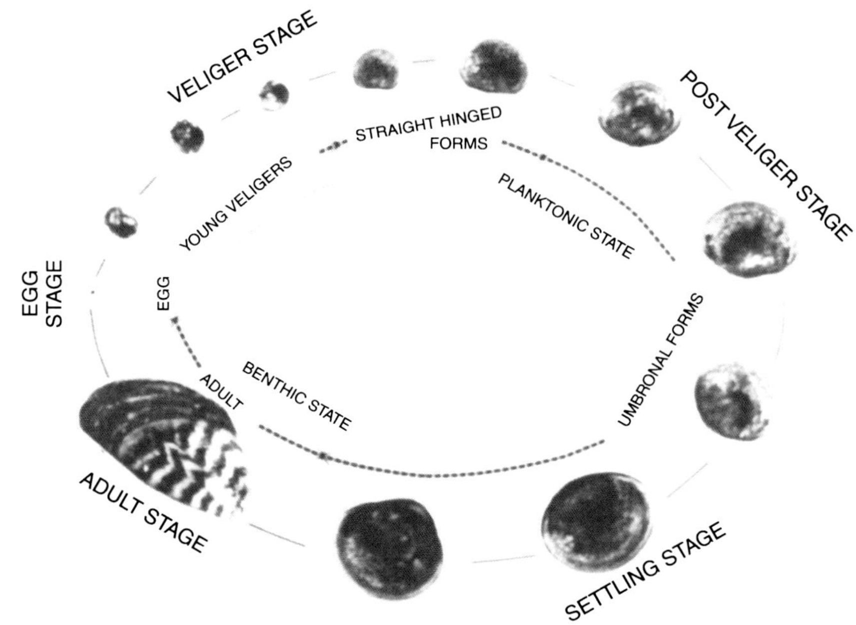

Source: G.L. Mackie, University of Guelph, Guelph, Ont.

Figure 9-1 Life cycle of the zebra mussel

Identification

Several features make the adult zebra mussel (shown in Figure 9-2) relatively easy to identify. The shell has a distinctly triangular shape, with one side of the triangle being broadly flattened, like a "D." A clump of byssal threads extends from the flattened side. The shell color ranges from tan or cream to dark brown, usually in alternating light and dark bands or zebra-like stripes.

The pattern and coloration of the shell striping is highly variable among specimens, as the species name *polymorpha* suggests. This variability may be more easily seen in the color photograph section. Zebra mussels may grow to 5 cm long, and masses of them may cover underwater structures completely, as can be seen in Figure 9-3.

Veliger larvae are typical of only two bivalve species, both of which have invaded North American freshwater environments and cause fouling problems. Besides the zebra mussel, the Asiatic clam, *Corbicula fluminea*, also produces a veliger larva. However, the zebra mussel, with its ability to attach and cluster, has proved to be a more severe problem as a biofouler of water supply systems.

The planktonic veliger larva appearing in Figure 9-4 is also relatively distinctive. The rotifer-size veliger displays the rudimentary appearance of a clam, but with a ciliated, bell-shaped velum (an organ enabling the mussel to swim) that extends outward. Veligers often swim with the velum extending upward. Disturbances in the water may cause the veliger to retract the velum and sink.

After hatching, the veliger is only 40 to 70 μm in diameter but quickly grows to 180 to 290 μm. After several weeks, the veliger undergoes a series of changes that distinguish it as postveliger. The most notable changes are that the velum is reduced in size and transformed into a siphon, and some organ systems develop. The animal eventually sinks and takes up a more sedentary form of existence.

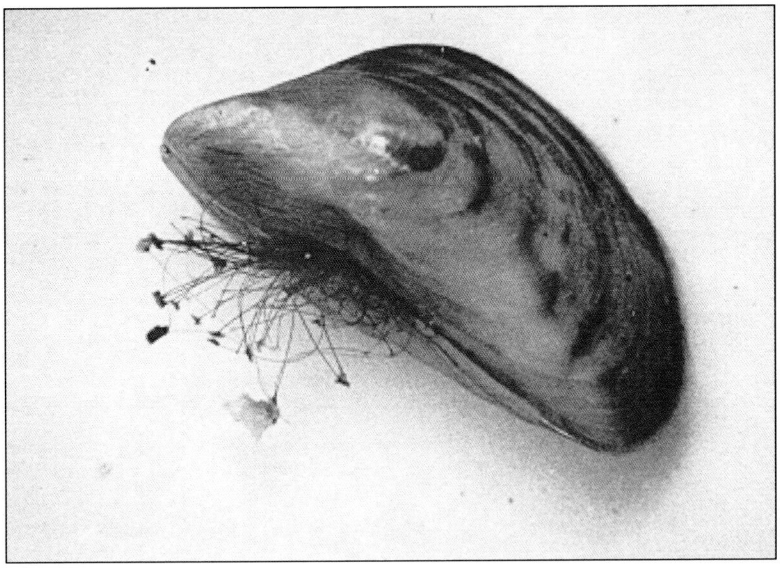

Source: F. Snyder, Ohio Sea Grant.

Figure 9-2 Adult zebra mussel

Source: Detroit Edison Company, Detroit, Mich.

Figure 9-3 Zebra mussels encrusted on structure

SIGNIFICANCE FOR WATER SUPPLIES

Zebra mussels' preference for surfaces submerged in flowing water at moderate depths has made water intake structures for power plants and municipal water systems particularly susceptible to colonization and clogging. As Figure 9-5 illustrates, mussels have caused significant reductions in water flow and pumping capability due to clogged screens, heat exchangers, and in-plant piping systems. Taste-and-odor problems associated with dead mussels have also been reported.

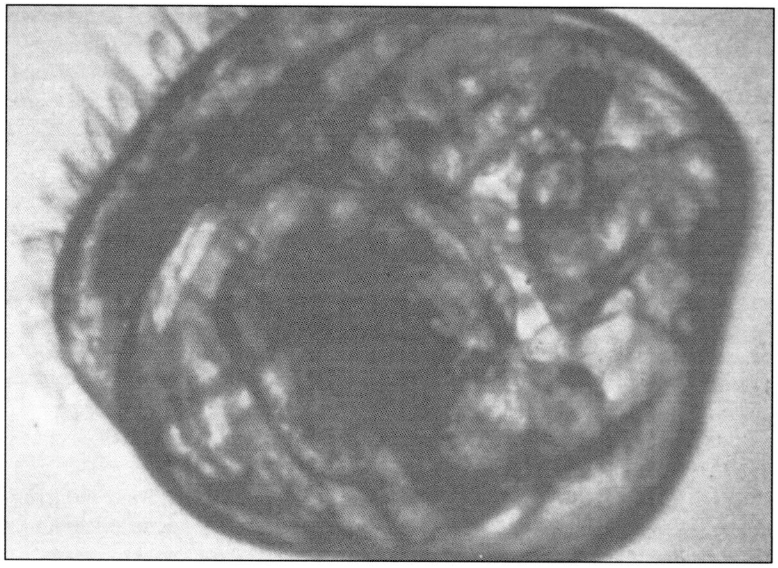

Source: T. Batterson, Great Lakes Sea Grant Network, Exotic Species Graphics Library.

Figure 9-4 Veliger larvae of the zebra mussel, showing the velum extended at lower right

Source: Detroit Edison Company, Detroit, Mich.

Figure 9-5 Impact of zebra mussels on screens

The zebra mussel may spread well beyond the waters of the Great Lakes, a result of both interbasin movement of commercial and recreational vessels, and possible transport by waterfowl and other aquatic animals. Its range may eventually encompass much of North America.

CONTROL STRATEGIES

Europeans have developed several effective zebra mussel control strategies. Researchers are assessing the applicability of European methods to North American conditions and are developing methods that are environmentally acceptable and cost-effective. Methods under consideration include chemical, physical, and mechanical controls.

Chemical Control

Chlorination at the water intake, either continuous or periodic, has proved effective in controlling zebra mussels, killing larvae and attached adults alike. Testing is currently under way to determine appropriate chlorine concentrations for use in North American waters. Environmental regulations generally require the dechlorination of water before discharge to the environment. High concentrations of chlorine also can limit the biological effectiveness of sand filters and cause taste-and-odor problems. Potassium permanganate can be effective as a predisinfectant for source waters with high organic content when the use of chlorine could cause trihalomethane formation problems. The addition of polymers at intake structures is showing promise in controlling zebra mussels.

Ozone, another oxidant, is being considered as an alternative to chlorine. Molluscicides, developed to treat other species, are being tested for effectiveness against the zebra mussel. However, any treatment that kills attached mussels will require the cleaning out and disposing of the dead organisms to avoid problems associated with clogging by detached shells and decaying mussels.

Antifouling coatings generally work by slowly releasing into the water a toxic substance, often an organometallic compound, that discourages the larvae from settling on the coated surface. These are generally expensive and have a short service life, which necessitates frequent reapplication. Negative environmental effects have caused a number of these compounds to be banned from use in fresh waters. Research is also being done on coatings that resist adhesion and colonization without the release of toxic products.

The use of any chemical control method may be subject to regulation by federal, state, provincial, and local agencies. To date, no chemical control has proved completely satisfactory.

Physical Control

Sand infiltration beds effectively prevent veliger larvae from entering water intake structures. However, this technique may be costly to put in place and maintain, and often results in reduced head pressure.

Thermal control has also proved effective. Water in the intake pipes and related structures must be raised to a temperature of at least 40°C and maintained for a certain period of time, generally over an hour. Water systems must be designed for this kind of treatment and the treatment may have to be repeated a number of times each season to be effective. The mussels that remain attached after treatment must be scraped off, and those that detach must be cleaned out as well.

Oxygen deprivation for a period of several days has also been used effectively in Europe. This method requires that the water intake system be shut down and sealed. Oxygen can be eliminated from the water in the system through the use of chemicals, or the system may be allowed to become anoxic naturally. Allowing the system to become anoxic on its own may require several weeks of shutdown. Dead mussels must be removed from the system for disposal through flushing and perhaps scraping.

The use of electrical fields to kill veligers and ultrasonic treatments to prevent settlement also are being considered as possible control methods.

Mechanical Control

Intake structures can be fitted with strainers, screens, or filters to block the entry of zebra mussels. Depending on mesh size, these may be effective against juvenile and adult zebra mussels, but veligers are likely to pass through any such structure.

Physically scraping mussels from the surfaces of the water system is also effective. This can be accomplished manually or through use of such devices as mechanical pigs. A variation on physical scraping is washing the affected surfaces with a fluid under high pressure to dislodge the mussels. Any such method requires the collection and disposal of dislodged zebra mussels, often in large quantities. System design and operational characteristics will determine the practicality and cost-effectiveness of this approach.

No one method of control will meet all needs. A combination of several of these methods often must be used for effective zebra mussel control. Control methods will also change as more research information becomes available, as new or improved methods come on line, and as facilities are upgraded. Choosing the most effective control strategy for a specific application requires the assistance of experienced professionals.

REFERENCES

Claudi, R., and G.L. Mackie. 1993. *Practical Manual for Zebra Mussel Monitoring and Control*. Boca Raton, Fla.: Lewis Publishers.

Electric Power Research Institute. *Zebra Mussel Monitoring and Control Guide: Final Report.* 1992. Palo Alto, Calif.: Electric Power Research Institute.

Hopkins, G. 1990. *The Zebra Mussel,* Dreissena polymorpha: *A Photographic Guide to the Identification of Microscopic Veligers.* Water Resources Branch, Limnology Section, Ontario Ministry of the Environment.

Mackie, G.L., W.N. Gibbons, B.W. Muncaster, and I.M. Gray. 1989. *The Zebra Mussel,* Dreissena polymorpha: *A Synthesis of European Experiences and a Preview for North America*. Water Resources Branch, Great Lakes Section, Ontario Ministry of the Environment.

Marsden, J.E. 1992. *Standard Protocols for Monitoring and Sampling Zebra Mussels.* Biological Notes #138, Center for Aquatic Ecology, Illinois Natural History Survey.

Nalepa, T.F., and D.W. Schloesser, eds. 1993. *Zebra Mussels: Biology, Impacts, and Control*. Boca Raton, Fla.: Lewis Publishers.

O'Neill, C.R. Jr., and D.B. MacNeil. 1989, revised 1990. Dreissena Polymorpha: *An Unwelcome New Great Lakes Invader.* New York: New York Sea Grant Institute.

Snyder, F.L. 1989. *Zebra Mussels in Lake Erie: The Invasion and Its Implications.* Fact Sheet No. 045, Ohio Sea Grant College Program.

Chapter 10

Algae

Algae can be defined as relatively simple, oxygen-producing, photosynthetic organisms, mostly microscopic in size. These organisms use light energy to convert carbon dioxide and water to sugars, and from these, to cell matter. Algae are a large and diverse group of organisms, exhibiting a great range of shapes, sizes, pigmentations, structural complexities, and life cycles.

BIOLOGY AND ECOLOGY

Algae range from bacterial-sized, single-cell forms to the giant marine kelps. They occupy a wide variety of habitats, including fresh water, oceans, estuaries, spray zones, moist soils, rice paddies, hot springs, stone or concrete surfaces, snowfields, and even deserts. Located in all climatic zones, from tropical regions to the arctic and antarctic, their great diversity in form and habitat makes algae difficult to define except in a general sense. Illustrations in the color section of this manual provide examples of algae that impact drinking water by causing taste-and-odor problems, clogging filters, and growing on reservoir walls. Examples are drawn from algae found in clean and polluted water and include planktonic and surface water algae. Found in both freshwater and marine environments, algae can exist in one of two general modes—planktonic or attached (also called sessile). Plankton are those organisms that are suspended in the water and are usually carried passively by the currents. Plankton can either be algal (phytoplankton) or animal (zooplankton). In contrast, periphyton are aquatic organisms that grow attached to some surface, such as submerged plants, rocks, the sediment surface, or reservoir walls. Those organisms that live on the bottom of a lake or reservoir are called benthic.

SIGNIFICANCE FOR WATER SUPPLIES

The significance of algae in water supplies arises from their mere presence in the water and from a variety of effects that they can exert on the aquatic environment. Tastes and odors are caused by the release of certain compounds by both living algae and dead and decomposing algae. This problem is common in water supplies in the United States and other parts of the world and will be discussed in greater detail in connection with the various algal groups. A review of odors produced by a wide variety of algae was published by Jüttner (1983), and the odorous compounds produced by eight different algae representing three major groups were determined by Rashash et al. (1996).

A variety of larger algae, when abundant in the water, can rapidly cover the surface of sand filters and drastically reduce the length of filter runs, thus necessitating frequent backwashings. The effective length of a filter run can be reduced from 50 hours or more to about 5 hours during a "bloom" (a proliferation of planktonic algae). In extreme cases, clogging may require more water to backwash than the amount of filtered water produced, greatly adding to the cost of treatment. This problem is usually caused by certain large diatoms, but various other algae can be responsible as well, especially those that form "flakes" or have sticky surfaces. Both filter-clogging and taste-and-odor algae are discussed and illustrated by Palmer (1957).

A massive bloom of planktonic algae can also cause rapid and severe oxygen depletion as algae die and decay, leading to fish kills. This phenomenon is associated primarily with blooms of certain colonial or filamentous blue–green algae that tend to form surface scums, but other algae may also form blooms. The floating algae not only exert an oxygen demand as they decay but also act as a physical barrier to the exchange of oxygen between the atmosphere and the water.

Another adverse effect associated with some algae is toxicity. A handful of blue–green algal species have toxic strains that have caused deaths of cattle, horses, hogs, waterfowl, and fish. Reports of blue–green algal poisonings have occurred in the United States, Canada, South Africa, Australia, and New Zealand, as well as other nations (Carmichael et al. 1985). There is some evidence that humans also can be affected by these toxins (Schwimmer and Schwimmer 1968; Hawkins et al. 1985), many of which have been isolated and their chemical structures determined (Gorham and Carmichael 1988). Some marine algae release a toxin that is taken up by shellfish and can lead to the serious condition called paralytic shellfish poisoning (PSP). Fortunately, this does not occur in freshwater sources.

Algae or their extracellular metabolites can also lead to the formation of trihalomethanes or other disinfection by-products when water containing the algae is chlorinated (Hoehn et al. 1980). These compounds are regulated in drinking water, and their concentrations must be minimized. Algae can also lead to elevation of pH that interferes with certain treatment processes. In some cases, pH levels as high as 9.5 have been observed in the upper depths of a reservoir experiencing an algal bloom. Finally, algae can cause obstruction in water conveyance systems, such as canals and aqueducts, where growth along the sides can impede the flow of water and require periodic physical removal or application of copper sulfate or chlorine. Extensive growth of some filamentous algae can also block and even damage screens and trash racks in reservoir outlet structures.

Algae can also have beneficial effects on water supplies. Because they release oxygen as part of their metabolism, they serve to oxygenate the water. Because blue–green algae are associated with water quality problems, a preponderance of algae other than blue–greens is desirable (especially green algae). Algae also play an important part in the aquatic food chain as they are the main food source for zooplankton and small fish, which in turn serve as food for larger fish and other wildlife. Algae are indicators of the trophic state of a water body; that is, the degree of pollution and nutrients in that water. A lake dominated by certain green algae and diatoms is a relatively "clean" oligotrophic water; whereas dominance by bloom-forming blue–greens indicates a more polluted or eutrophic condition.

Algae are useful in wastewater treatment, supplying oxygen in oxidation ponds for aerobic bacteria, fungi, and other microorganisms that break down organics in wastewater. This organic fraction constitutes a large part of the biological oxygen demand (BOD) in sewage and must be removed before discharge. Algae also take up (and thereby remove) various minerals, including phosphorus, ammonium, calcium, magnesium, potassium, and some heavy metals (Oswald 1988). Removal of these constituents is important for ensuring water quality in receiving streams, lakes, or rivers. In addition, some blue–green algae able to assimilate atmospheric nitrogen (N_2) contribute to the fertility of soils in rice fields. This is very important in some Asian countries where rice is a staple of their diet.

ALGAL DIVISIONS

There are eight to eleven phyla or divisions of algae (depending on the taxonomy) that represent fundamental differences between the divisions. The descriptions of the major algal phyla include the most important characteristics visible with the light microscope but do not include every structural or biochemical feature that may distinguish one from another. Only freshwater algae will be considered in this chapter. The Phaeophyta, or brown algae, are omitted as they are almost exclusively marine. Another group, the Prymnesiophytes, was omitted because it is found mostly in brackish water. Details of the various algal groups and the genera that comprise them can be found in any general phycology text, such as those referenced under Prescott (1962), Bold and Wynne (1985), and Wehr and Sheath (2003). An excellent book with many color photographs is Carter-Lund and Lund (1995). The general characteristics separating algal groups are morphology, pigments, storage products, cell walls, motility, and reproduction. Many of these properties are discussed below.

Chlorophyta (Green Algae)

General description. The Chlorophyta, or green algae, constitute by far the largest and most diverse group of freshwater algae in terms of numbers of species, range of structural organization, and frequency of occurrence. Green algae can be single-celled, colonial, filamentous, membranous, and tubular. They exhibit a wide range of life cycles. Generally, the chlorophytes are a medium- to grass-green color when healthy, and contain the pigments chlorophyll *a* and *b*, alpha- and beta-carotenes (orange), and several xanthophylls (yellow). The chlorophylls are responsible for photosynthesis. The various other algal pigments can have either light-harvesting functions or protective functions for the cells. The food storage product is starch. The cellular organization, as in all algae except blue–greens, is eucaryotic, with a distinct nucleus and other units or organelles with particular functions. Chlorophyll is contained in an organelle called a chloroplast, which is the site of photosynthesis and the most conspicuous part of the cell. The cell wall is composed of cellulose and pectose (a sugar). Many green algae have swimming cells with flagella (tail-like extensions) as part of the life cycle.

Some of the more common and well-known genera are *Actinastrum*, *Chlamydomonas*, *Chlorella*, *Coelastrum*, *Cladophora*, *Hydrodictyon*, *Oocystis*, *Pediastrum*, *Scenedesmus* (shown in color section), *Spirogyra*, and *Volvox*.

Occurrence. Green algae occur in a wide range of freshwater and marine habitats and can be both planktonic and benthic. They are very common in lakes, ponds, ditches, wet soil, and even in treated water reservoirs. Some species occur in soft, acid waters, which are generally unsuitable for most other algae.

Significance in water supplies. Green algae are generally considered desirable, or at least harmless, in water supplies, although they can sometimes proliferate in treated water reservoirs, causing "green water" to flow out into the distribution system. Some, such as *Pandorina*, *Chlamydomonas*, and *Volvox*, can cause taste-and-odor problems when abundant in the water, but this is usually not severe. Some filamentous green algae, such as *Cladophora* and *Rhizoclonium*, can form very conspicuous wiry growths and tangled mats in ponds or flowing water on dams.

Cyanophyta (Blue–Green Algae)

General description. Blue–green algae are fundamentally different from all other algal divisions in that their level of cellular organization is "procaryotic," that is, relatively simple and uncompartmentalized, more similar to bacteria than to algae. This affinity to bacteria is acknowledged in their more modern name, cyanobacteria. There is no distinct nucleus as in higher organisms, nor other organelles, such as chloroplasts. Some species contain gas vacuoles, used for buoyancy regulation in the water. Blue–green algae can be single-celled, colonial, or filamentous, either simple or branched. They contain chlorophyll *a* (green), phycoerythrin (red), phycocyanin (blue), and various other pigments.

Color may range from dark green or blue–green to pink, red, lavender, brown, or black. Color alone cannot be used to distinguish cyanophyta from other algae. There is no sexual reproduction; reproduction is by cell division or by spores (endospores or akinetes). Some blue–greens can move by a gliding motion over solid surfaces. Certain species are capable of nitrogen fixation, a process whereby N_2 from the atmosphere, which is generally unavailable to most organisms, is reduced and converted to a form available to the organisms for growth. This process occurs in a structure called a heterocyte (formerly called heterocyst).

Some of the more prominent genera in water supplies are *Anabaena*, *Aphanizomenon*, *Gomphosphaeria*, *Gleotrichia*, *Microcystis*, *Oscillatoria*, and *Phormidium*. Several excellent books on this algal group have been written by Carr and Whitton (1982), Fay and Van Baalen (1987), and Whitton and Potts (2000).

Occurrence. Blue–green algae occupy many different habitats, ranging from lakes or reservoirs (where the algae can be either planktonic or benthic), streams, wet soils, irrigation ditches, stone or concrete surfaces in tropical areas, alkaline hot springs (such as found in Yellowstone Park), snowfields, and even deserts. They also occur in salt water and estuaries. Some exist in a symbiotic association with certain fungi (as lichens) or inside some plants or protozoa. Blue–green algae are believed to have been the first oxygen-evolving photosynthetic organisms on earth, having been found in rock formations about 2.6 billion years old, and are considered to have been responsible for the development of an oxygen atmosphere that made possible all the higher forms of life (Schopf 1983).

Significance in water supplies. Most of the effects discussed under "significance of algae" are attributable to blue–green algae; namely, taste and odor, filter clogging, oxygen depletion, and toxicity. There are many species that produce the earthy–musty compounds geosmin and 2-methylisoborneol (MIB). These compounds are detectable by many people at extremely low concentrations (10 ng/L or less) and are difficult to remove except by ozone or granular activated carbon. Most of these odor-producing species belong to the genera *Anabaena*, *Aphanizomenon*, *Lyngbya*, *Oscillatoria*, *Phormidium*, or *Pseudanabaena* and have been identified as taste-and-odor producers only within the last 20 years (Tabacheck and Yurkowski 1976; Izaguirre et al. 1982; Izaguirre and Taylor 1998). In addition, some major bloom-formers, such as *Microcystis*, release a variety of odorous organic sulfur compounds, especially when they decay. Several reviews on odor production by cyanophyta are found in Slater and Blok (1983) and in Mallevialle and Suffet (1987), with an update in chapter 2 of Suffet, Mallevialle and Kawczynski (1995). Some planktonic blue–greens tend to float to the surface of the water during warm, calm weather and form scums, which are unsightly and foul-smelling when the algae decompose. The planktonic blue–greens can also cause severe oxygen depletion, primarily from decay of dead algae. This problem is especially undesirable in lakes or reservoirs that have recreational uses in addition to providing a source of drinking water. It is important to remember that not only blue–green algae produce "blooms." A source water may experience a spring bloom of *Synedra* followed by a fall bloom of *Tabellaria*, for example. These diatoms are then consumed by *Daphnia*, which also clog filters.

An area receiving increased attention during the last two decades is toxin production by certain blue–green algae. The species *Anabaena flos-aquae*, *Aphanizomenon flos-aquae*, *Microcystis aeruginosa, Cylindrospermopsis raciborskii,* and *Nodularia spumigena* have toxic strains that can coexist with nontoxic strains in the same bloom. The toxic strains either release the toxins into the water (especially on death), or in the case of *Microcystis*, bind them up in the cells. Some of these toxins are lethal when injected into mice or rats. The toxins of *Anabaena* and *Aphanizomenon* are primarily neurotoxins, and those of *Microcystis, Cylindrospermopsis* and *Nodularia* are hepatotoxins (affecting the liver). These compounds have been found in certain other species as well but less frequently.

Most blue–green algal poisonings have involved livestock or waterfowl, but there is some evidence that humans can be affected as well, especially if they ingest water from a dense bloom (Hawkins et al. 1985). Until recently, there were no reports of fatal human poisonings by toxic freshwater blue–green algae. However, in the 1990s there were two reports of human deaths in Brazil resulting from algal toxins in water (Texeira et al. 1993; Jochimsen et al. 1998), one of them involving patients in a hemodialysis center. These incidents showed that under certain conditions, algal toxins can be hazardous to human health. The main reason that people are not more often affected is probably that they are repelled by the sight of a thick algal bloom in a lake and will not willingly drink the water. A second reason is that the toxic algae often accumulate at or near the surface of lakes and reservoirs and are therefore not abundant in the depths from which water is usually withdrawn. The toxins can best be removed from water by activated carbon (Falconer et al. 1989) and ozone (Chorus and Bartram 1999, ch. 9). Several excellent reviews on algal toxins have been published recently (Gorham and Carmichael 1988; Chorus and Bartram 1999). The results of a survey on the occurrence of microcystins in many water utilities in the United States and Canada were reported by Carmichael (2001).

Chrysophyta (Yellow–Green or Golden-Brown Algae)

General description. The Chrysophyta comprise six classes, only three of which are significant in water supplies. This group of algae is quite diverse with respect to pigment composition, cell wall, and type of flagellated cells, yet they share certain features. Their name is derived from the predominance of carotenoids (orange–yellow pigments) over chlorophylls (green), hence their name. Another common feature is the type of food reserve (chrysolaminaran). Most classes of this division have chlorophyll *a* and *c*, and none have chlorophyll *b*.

Main classes. The following are the three most important chrysophyta classes for freshwater supplies.

Chrysophyceae. There is a great deal of variability in both morphology and mode of nutrition in this algal class. Although many chrysophycean algae do not have a cell wall, others have various cell coverings, including scales, loricas (a type of encasing), and cell walls. Some species may lose their chloroplasts and become colorless. One of the most distinctive features of this class is the formation of a characteristic cyst, or statospore, that constitutes a resting stage formed inside the cell. Many members of this class have flagella, usually two.

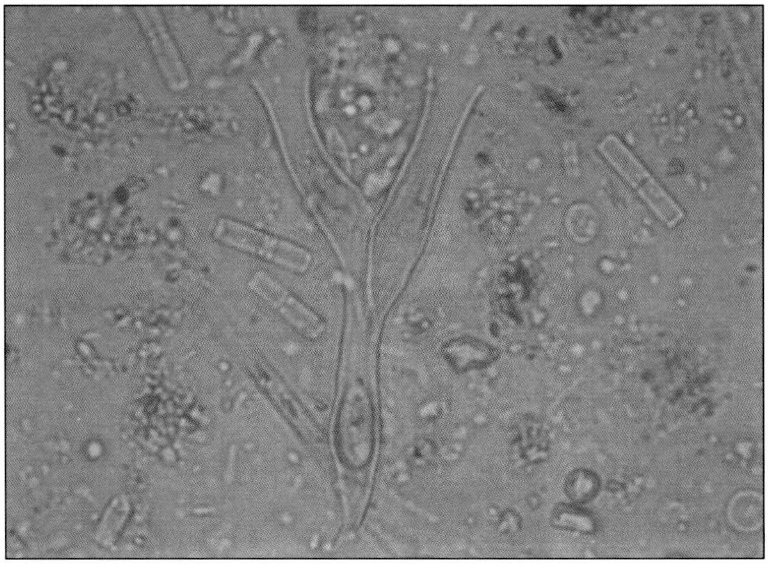

Source: K. Hancock, CHDiagnostic & Consulting Service Inc.

Figure 10-1 Algae, Chrysophyte, *Dinobryon* (3 cells), and others, unstained bright-field, 240×

Some of the more common and important chrysophycean algae in drinking water supplies are *Dinobryon* (shown in Figure 10-1 and in the color photograph section), *Mallomonas*, *Synura*, *Uroglena*, and *Uroglenopsis*.

Xanthophyceae. The Xanthophyceae (also called Tribophytes) differ from the Chrysophyceae and the diatoms (discussed on page 63) by the lack of the pigment fucoxanthin, the absence of which causes these algae to appear yellow–green rather than the golden or brownish color of the other classes. Members of this class are easily confused with green algae because of their greenish to yellow–green appearance. Testing for starch (by adding iodine) would result in a negative reaction, indicating that the algae in question is xanthophycean rather than chlorophycean. Some of the more common genera are *Tribonema* and *Vaucheria*.

Bacillariophyceae (Diatoms). Diatoms are a very important algal class in terms of both number of species and widespread distribution. It is rare when an examination of a sample of lake or pond water does not reveal some diatoms. There are 58 freshwater genera, many more marine or brackish water genera, and numerous extinct fossil forms. All diatoms, despite the great variability in form, share the common feature of having a rigid cell wall composed of silica with an organic coating, called a frustule.

The classification of diatoms is almost entirely based on the structure and ornamentation of the frustule. Diatoms are unusual among microorganisms in that it is often easier to examine and identify them using nonliving specimens, specially cleaned and prepared, than living specimens. The frustule consists of two overlapping halves, the two surfaces being the valves. Located between the two valves is the girdle. Diatoms have two orientations as seen by the observer—a valve view and a girdle view—which can appear quite different.

Descriptions of diatoms usually mention the shape of both the valve and girdle view. Various types of markings on the valve surfaces may be present, including "dots," pores, and "ribs." The frustules (shells) can be either round, or boat- or cigar-shaped. The chloroplasts are usually a conspicuous part of the cell and are usually golden brown, but may also be yellowish-green or dark brown. Some of the most common genera are *Asterionella* (shown in Figure 10-2 and in the color photograph section), *Cyclotella*,

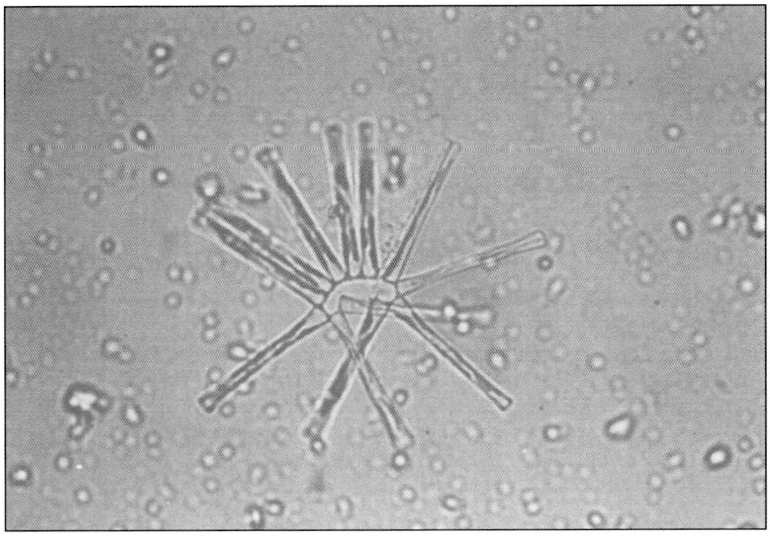

Source: K. Hancock, CHDiagnostic & Consulting Service Inc.

Figure 10-2 Algae, Diatom, *Asterionella*, unstained bright-field microscopy, 120×

Fragilaria (also in the color section), *Gyrosigma*, *Melosira*, *Nitzchia*, *Navicula*, *Stephanodiscus*, *Synedra*, and *Tabellaria*.

Occurrence. Chrysophyta occur in a wide variety of habitats. Most of the Chrysophyceae are planktonic and are found in lakes, reservoirs, ponds, or mountain streams. They are most abundant in lakes during spring and autumn when the water is cool. The Xanthophyceae include planktonic and attached forms, and some that grow on damp walls or soil, or intermingled with mosses or lichens. Diatoms are ubiquitous in their distribution and include planktonic and attached species, and many that grow on moist soil. Attached forms may grow on lake or reservoir bottoms (benthic) or attached to plants (epiphytic). They are very common in reservoir sediments and even grow on the walls of filters in treatment plants. Spring diatom blooms are common in lakes in the United States.

Significance for water supplies. Chrysophyceae and diatoms include a number of species notorious for the taste and odor that they impart to drinking water. Among the former class, *Dinobryon* (seen in Figure 10-1), *Mallomonas*, *Synura*, *Uroglena*, and *Uroglenopsis* produce either a "fishy" or "cod-liver oil" odor and flavor. The odor of some *Synura* species has been described as "cucumber-like." Some of the compounds responsible have been identified (Jüttner 1981; Yano, Nakahara, and Ito 1988; Rashash et al. 1996) Some diatoms are also associated with taste-and-odor problems, with the aroma usually characterized as fishy or "oily." *Cyclotella* and *Melosira* are two examples. *Asterionella* causes a taste more than an odor.

Various diatom species are important as filter cloggers when present in high numbers, notably species of *Synedra*, *Asterionella*, *Fragilaria*, *Melosira*, and *Tabellaria* (Baylis 1955). Some of these diatoms, such as *Fragilaria*, *Melosira*, and *Tabellaria*, grow as chains of cells. *Asterionella* occurs as six or more cells radiating from a central point like spokes of a wheel (Figure 10-2). These configurations make them more likely to collect on the surface of a filter and eventually clog it. Some diatoms also tend to grow on reservoir walls, such as the genera *Achnanthes*, *Gomphonema*, and *Cymbella*. The first two grow at the ends of stalks, and the third grows inside hollow tubes, resulting in unsightly growths that can break off and clog screens and filters.

Pyrrhophyta (Dinoflagellates)

General description. Dinoflagellates constitute an important component of plankton in oceans, brackish water, and fresh waters. In most freshwater forms, cells are solitary and mobile. Swimming is most often accomplished by two flagella of approximately equal length that are usually attached on the side. The arrangement of the flagella causes the cells to rotate as they swim. The cell wall may be firm and simple, or formed of regularly arranged, polygonal plates. Pigments include chlorophyll *a* and *c*, beta-carotene (orange), and various xanthophylls (yellow pigments), peridinin and dinoxanthin being most abundant. Reproduction is usually by longitudinal division. Some of the more commonly seen freshwater dinoflagellates are *Peridinium*, *Ceratium*, and *Glenodinium*.

Occurrence. Most dinoflagellates are found in marine plankton, especially in warmer parts of the world. Freshwater dinoflagellates are most abundant in pools, ditches, and small lakes with plentiful vegetation. Some species thrive best in hard waters, while others are most abundant in soft waters. Dinoflagellates are best known for the ocean blooms called "red tides." Some of these blooms are associated with toxins that cause PSP.

Significance for water supplies. The main problem that dinoflagellates can cause in water supplies is taste and odor, with *Ceratium* and *Peridinium* foremost in this regard. Both of these genera produce "fishy" odors, and *Ceratium* purportedly causes "septic" odors when the organisms are abundant or have died.

Euglenophyta

General description. Euglenoids are intermediate in some respects between algae and protozoa and are even included among protozoa in some textbooks. Cells are solitary, without a cell wall, and swim by means of one or two (rarely three) flagella. A cavity containing air or fluid is present at the front end of the cell, as well as a red eyespot. The cell membrane is in the form of a layer composed of overlapping strips that may be rigid or flexible. Pigments include the chlorophylls *a* and *b*, beta-carotenes, and several xanthophylls. Chloroplasts are few to many and may be small, simple discs; large and plate-like; or ribbon-like and arranged in star-shapes. Some forms are colorless. The nucleus is very conspicuous, and reproduction is by longitudinal cell division.

Occurrence. Euglenoids are widely distributed, occurring in fresh water, brackish and marine waters, and on moist soils and mud along riverbanks. They are often abundant in ponds rich in organic matter, and in tanks and puddles, especially those to which livestock have access. Euglenoids often occur in sufficient numbers to color the water green.

Significance for water supplies. Euglenoids are generally indicative of polluted water, including pollution by domestic sewage or effluent. Three species are particularly important as indicators of pollution—*Euglena viridis*, *Lepocinclis texta*, and *Phacus pyrum*—although other species of these genera may also be associated with this problem (Palmer 1957). A photograph of *Phacus* appears in the color section. Euglenoids are found in streams subject to organic enrichment and usually occur with a variety of other algae that comprise a recognizable community. One euglenoid, *Trachelomonas crebea*, is sometimes a filter clogger.

Cryptophyta (Cryptomonads)

General description. Cryptophyta or cryptomonads are a relatively small group of organisms whose asymmetrical cells are flattened from top to bottom and bounded by a periplast. A periplast consists of a cell membrane with an underlying layer of plates or membranes and an overlying layer of granular material. Cells combine a degree of firmness with some flexibility. Two flagella always arise ventrally from within a depression or furrow, the opening of which is close to the front of the cell. Flagella may be equal or unequal in length.

A broad range of pigmentation is evident in the cryptomonads, and some colorless forms also exist. Cells may be red, blue, olive-yellow, green, or brown. Pigment may change with age, making color unreliable as a trait for identification, particularly at the generic level. Usually there is only one or a pair of chloroplasts in a cell. The eyespot is usually in the middle of the cell, close to the nucleus. Photosynthetic pigments include chlorophyll *a* and *c*, alpha- and beta-carotene, and some xanthophylls. Other pigments are also present that are responsible for the reddish to bluish hues seen in many species. Reproduction is by longitudinal cell division.

Occurrence. Members of this group occur in both freshwater and marine habitats. Freshwater forms are very common in lakes and are often the dominant algae in the plankton. Cryptomonads are more commonly associated with shallow waters around decaying vegetation or in ponds and stagnant waters.

Significance for water supplies. Cryptomonads are of minor significance in water supplies. Some genera, such as *Cryptomonas*, may cause taste and odor when abundant.

Rhodophyta (Red Algae)

General description. The majority of Rhodophyta, or red algae, are marine, but there are about 15 freshwater genera in this division. Red algae have the following characteristics: absence of any flagella stages, the presence of a different type of photosynthetic pigment called phycobilins (phycoerythrin and phycocyanin, also found in Cyanophyta), and a highly specialized manner of sexual reproduction. Phycoerythrin, usually the predominant pigment, is responsible for the red color in these plants, often masking the presence of chlorophyll *a*. Freshwater forms are usually colors other than red, such as violet, olive-green, and brown. The color also depends on their depth in the water.

The great majority of Rhodophyta are filamentous, foliose (lobed or leaf-like), or more massive forms. There are also a few genera of single-celled red algae as well as simple filamentous and colonial forms. In the larger and more complex forms, the plant body is called a thallus. Many species attach themselves to the substrate by various means, including multicellular holdfasts. Some red algae have the capacity to deposit calcium carbonate in the cell wall. The reproductive cycles of red algae are relatively complex and beyond the scope of this manual. The interested reader is referred to a phycology text, such as Bold and Wynne (1985), or Wehr and Sheath (2003).

Occurrence. Freshwater red algae constitute an insignificant portion of a division that is primarily marine. Freshwater genera are usually without representatives in the ocean. The great majority of freshwater forms are restricted to the well-aerated waters of rapids, falls, and dams in cold, rapidly flowing streams. However, a few genera are found in quiet and relatively warm waters.

Significance for water supplies. Red algae have minor impact on drinking water supplies. Three species, *Audouinella violacea*, *Batrachospermum moniliforme*, and *Compsopogon coeruleus*, can grow on reservoir walls. *Compsopogon* has also been found in irrigation ditches in Arizona.

CONTROL STRATEGIES

Copper sulfate ($CuSO_4$), long used for algae control (Moore and Kellerman 1905), remains the algicide of choice in potable water supplies. The application of $CuSO_4$ to lakes, reservoirs, or rivers can be an important and effective management tool but ideally should be used in conjunction with other techniques. The amount of $CuSO_4$ needed in any situation depends on the chemistry of the water, particularly pH and alkalinity, and on the sensitivities of the algae themselves. Attached algae require larger crystals of $CuSO_4$ than planktonic algae. For plankton, $CuSO_4$ is usually applied as a fine crystalline form, or "snow." Application methods can vary from the traditional burlap bag towed behind a boat to mechanical spreaders, sprayers, and helicopters. General guidelines for $CuSO_4$ doses required to control phytoplankton, developed by Mackenthun (1961), are as follows:

- For lakes with a methyl orange alkalinity above 40 mg/L as calcium carbonate, the dose should be 0.58 g/m^2 (5.4 lb/acre) of $CuSO_4 \cdot 5H_2O$ (equal to 1.0 mg/L) for the top 2 ft (0.6 m)
- For lakes with a lower alkalinity, the applied dose should be 0.3 mg/L (0.9 lb/acre)

There are apparently no general guidelines for control of periphyton, except that $CuSO_4$ should be applied as crystals or chunks large enough to sink to the bottom. Copper sulfate generally is applied only to the shallow areas of the reservoir because it becomes diluted and less effective in deeper waters. Several case histories appear in McGuire et al. (1984) and Casitas Municipal Water District (1987). When $CuSO_4$ is used, it should be accompanied by a monitoring program to determine effectiveness and to minimize excessive chemical usage. Excessive copper becomes incorporated in sediments where it can be toxic to benthic organisms and plants that are essential to the health of the reservoir.

Other methods for controlling algae include chlorination, artificial destratification, control of nutrient input, and food web management. These approaches are discussed in detail in an excellent review of reservoir management techniques by Cooke and Carlson (1989), and in an AWWA report on algal control methodologies (Casitas Municipal Water District 1987). Both of these books contain many case histories of the various methods of control for algae. Other approaches are $Ca(OH)_2$ addition to coagulate algal cells or precipitate phosphorus (Zhang and Prepas 1996) or the use of barley straw to inhibit planktonic blue–green algae (Everall and Lees 1996). Removal of algal cells in treatment plants has been achieved with dissolved air flotation (Markham et al. 1997).

REFERENCES

American Public Health Association, American Water Works Association, Water Environment Federation. 1998. *Standard Methods for the Examination of Water and Wastewater.* Latest edition. Washington, D.C.: American Public Health Association.

Baylis, J.R. 1955. Effect of Microorganisms on Lengths of Filter Runs. *Water Wks. Eng.*, 108:127–128,158.

Bold, H.C., and M.J. Wynne. 1985. *Introduction to the Algae: Structure and Reproduction.* 2nd ed. Englewood Cliffs, N.J.: Prentice Hall Inc.

Carmichael, W.W., C.L.A. Jones, N.A. Mahmood, and W.C. Theiss. 1985. Algal Toxins and Water-Based Diseases. *CRC Crit. Rev. in Environ. Contr.*, 15:275–313.

Carmichael, W.W. 2001. *Assessment of Blue-Green Algal Toxins in Raw And Finished Drinking Water.* AWWA Research Foundation and American Water Works Association, Denver, CO.

Canter-Lund, H., and J.W.G. Lund. 1995. *Freshwater Algae: Their Microscopic World Explored*. Biopress Ltd., Bristol, England.

Carr, N.G., and B.A. Whitton, eds. 1982. *The Biology of Cyanobacteria*. Oxford, England: Blackwell Scientific Publishing.

Casitas Municipal Water District. 1987. *Current Methodology for the Control of Algae in Surface Reservoirs*. Denver, Colo.: American Water Works Association Research Foundation and American Water Works Association.

Chorus, I., and J. Bartram, eds. 1999. *Toxic Cyanobacteria in Water: a Guide to Their Public Health Consequences, Monitoring and Management*. E. & F.N. Spon, London and New York.

Cooke, G.D., and R.E. Carlson. 1989. *Reservoir Management for Water Quality and THM Control*. Denver, Colo.: American Water Works Association Research Foundation and American Water Works Association.

Everall, N.C., and D.R. Lees. 1996. The Use of Barley-Straw to Control General and Blue-Green Algal Growth in a Derbyshire Reservoir. *Wat. Res.*, 30:2:269–276.

Falconer, I.R., M.T.C. Runnegar, T. Buckley, V.L. Huyn, and P. Bradshaw. 1989. Using Activated Carbon to Remove Toxicity From Drinking Water Containing Cyanobacterial Blooms. *Jour. AWWA*, 81:2:102–105.

Fay, P., and C. Van Baalen, eds. 1987. *The Cyanobacteria*. Elsevier Science Publishers, Amsterdam, The Netherlands.

Gorham, P.R., and W.W. Carmichael. 1988. Hazards of Freshwater Blue-Green Algae (Cyanobacteria). *Algae and Human Affairs*. pp. 403–431. C.A. Lembi and J.R. Waaland, eds. Cambridge, England: Cambridge University Press.

Hawkins, P.R., M.T.C. Runnegar, A.R.B. Jackson, and I.R. Falconer. 1985. Severe Hepatotoxicity Caused by the Tropical Cyanobacterium (Blue-Green Alga) *Cylindrospermopsis raciborskii* (Wolosz.) Seenaya and Subba Raju Isolated From a Domestic Water Supply Reservoir. *Applied and Environmental Microbiology*, 50:1292–1295.

Hoehn, R.C., D.B. Barnes, B.C. Thompson, C.W. Randall, T.J. Grizzard, and P.T.B. Shaffer. 1980. Algae as Sources of Trihalomethane Precursors. *Jour. AWWA*, 72:344–350.

Izaguirre, G., C.J. Hwang, S.W. Krasner, and M.J. McGuire. 1982. Geosmin and 2-methylisoborneol From Cyanobacteria in Three Water Supply Systems. *Applied and Environmental Microbiology*, 43:708–714.

Izaguirre, G., and W.D. Taylor. 1998. A *Pseudanabaena* Species from Castaic Lake, California that Produces 2-Methylisoborneol. *Wat. Res.*, 32:5:1673–1677.

Jüttner, F. 1981. Detection of Lipid Degradation Products in the Water of a Reservoir During a Bloom of *Synura uvella*. *Applied and Environmental Microbiology*, 41:100–106.

———. 1983. Volatile Odorous Excretion Products of Algae and Their Occurrence in the Natural Aquatic Environment. *Water Science and Technology*, 15:247–257.

Mackenthun, K.M. 1961. The Practical Use of Present Algicides and Modern Trends Toward New Ones. *Algae and Metropolitan Wastes*. Trans. of 1960 seminar, US Dept. Health, Education, and Welfare, US Public Health Service. PB-199-296. Cincinnati, Ohio.

———. 1971. Control of Biological Problems in Water Supplies. Denver, Colo.: Education committee of AWWA and USEPA, Water Hygiene Division–Water Quality Office.

Mallevialle, J., and I.H. Suffet, eds. 1987. *Identification and Treatment of Tastes and Odors in Drinking Water*. pp. 53–101. Denver, Colo.: American Water Works Association Research Foundation and Lyonnaise des Eaux, France.

Markham, L., M. Porter, and T. Schofield. 1997. Algal and Zooplankton Removal by Dissolved Air Flotation at Severn Trent Ltd. Surface Water Treatment Works. In *Dissolved Air Flotation*. Proceedings of an International Conference, Charted Institution of Water and Environmental Management, London.

McGuire, M.J., R.M. Jones, E.G. Means, G. Izaguirre, and A.E. Preston. 1984. Controlling Attached Blue-Green Algae with Copper Sulfate. *Jour. AWWA*, 76:5:60–65.

Moore, G.T., and K.F. Kellerman. 1905. Copper as an Algicide and Disinfectant in Water Supplies. *Bull. Bureau Plant Indus.* 76, pp. 19–55. US Dept. of Agriculture.

Oswald, W.J. 1988. The Role of Microalgae in Liquid Waste Treatment and Reclamation. In *Algae and Human Affairs*. pp. 255–281. C.A. Lembi and J.R. Waaland, eds. Cambridge England: Cambridge University Press.

Palmer, C.M. 1957. Algae as Biological Indicators of Pollution. *Biological Problems in Water Pollution*. pp. 60–69. US Dept. Health, Education, and Welfare, US Public Health Service, Robert A. Taft San. Eng. Center. Cincinnati, Ohio.

Prescott, G.W. 1962. *Algae of the Western Great Lakes Area*. Dubuque, Iowa: Wm. C. Brown Company Publishers.

Schopf, J.W., ed. 1983. *Earth's Earliest Biosphere. Its Origin and Evolution*. Princeton, N.J.: Princeton University Press.

Schwimmer, M., and D. Schwimmer. 1968. Medical Aspects of Phycology. *Algae, Man, and the Environment*. pp. 278–358. D.F. Jackson, ed. Syracuse, N.Y.: Syracuse University Press.

Slater, G.P., and V.C. Blok. 1983. Volatile Compounds of the Cyanophyceae—A Review. *Water Science and Technology*, 15:181–190.

Suffet, I.H., J. Mallevialle, and E. Kawczynski, eds. 1995. *Advances in Taste-and-Odor Treatment and Control*. AWWA Research Foundation and Lyonnaise des Eaux, France.

Tabacheck, J.L., and M. Yurkowski. 1976. Isolation and Identification of Blue-Green Algae Producing Muddy Odor Metabolites, Geosmin and 2-Methylisoborneol, in Saline Lakes in Manitoba. *J. Fish. Res. Board Can.*, 33:25–35.

Wehr, J.D., and R.G. Sheath, eds. 2003. *Freshwater Algae of North America*. Academic Press, New York.

Whitton, B.A., and M. Potts, eds. 2000. *Ecology of Cyanobacteria: Their Diversity in Time and Space*. Kluwer Academic Publishers, Dordrecht, The Netherlands.

Yano, H., M. Nakahara, and H. Ito. 1988. Water Blooms of *Uroglena americana* and the Identification of Odorous Compounds. *Water Science and Technology*, 20:75–80.

Zhang, Y., and E.E. Prepas. 1996. Short-Term Effects of $Ca(OH)_2$ Additions on Phytoplankton Biomass: a Comparison of Laboratory and *In Situ* Experiments. *Wat. Res.*, 30:5:1285–1294.

Chapter 11

Protozoa

Protozoa are unicellular animals that, unlike bacteria and viruses, possess membrane-bound genetic material or nuclei and other assorted cellular organelles. Their methods of locomotion and reproduction are used to categorize them into broad groups. However, there is little agreement about the taxonomy of the protozoa, which is a dynamic, evolving process. Protozoa fall into two broad groupings depending on whether they are parasitic or free-living.

Free-living protozoans are widespread in natural waters as well as moist soils and fall into three broad taxonomic groups—amoebae, flagellates, and ciliates. Examples of these three groups may be found in the color section of this manual. They are often abundant in both surface and groundwater supplies as part of the normal aquatic community. Few of these protozoans will come to the attention of water treatment plant operators, because few free-living species will cause treatment problems or health impacts. However, some free-living protozoa will cause disease and death. Operators should be aware that biologically active processes in water treatment, such as slow sand filters and biofilms, in which large numbers of bacteria are present, will most likely have active protozoan communities.

In groundwaters, amoebae are frequently observed in samples submitted for microscopic particulate analysis (MPA) for surface water under the influence of groundwater. Testate amoebae, such as *Arcella* (illustrated in Figure 11-1), *Cryptodifflugia*, and *Quadrulella*, are found in association with a variety of environmental bacteria, including the iron bacteria. While groundwaters probably are populated by many other protozoa, the method of sample collection for MPA destroys more fragile members of the groundwater protozoan community.

Identification of free-living protozoa to genus and species requires a great deal of skill and training. Keys and pictures like those in *An Illustrated Guide to the Protozoa, 2nd edition* (Lee, Leedale, and Bradbury 2002) aid in identifying unknown organisms.

A few of the free-living protozoa may cause disease and death. Periodically in North America, one or more swimmers inhale pathogenic protozoa, such as *Acanthamoeba* and *Naegleria fowleri*, and die of meningitis. However, no waterborne disease outbreak has been associated with these organisms in North America. Included as representatives of free-living protozoa are *Hartmanella* spp., *Acanthamoeba* spp., *Echinamoeba* spp.,

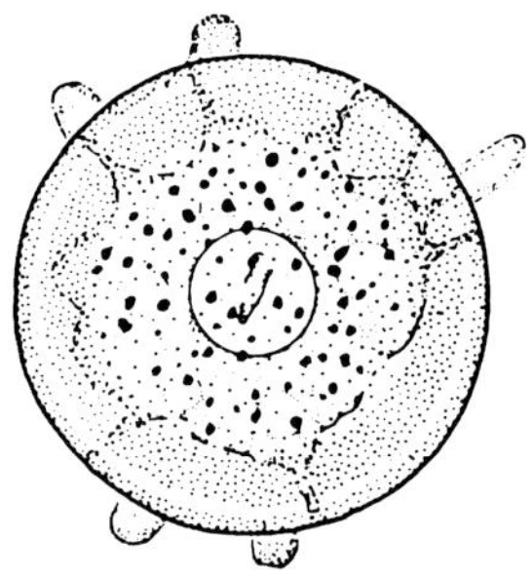

Source: N. Strebel and D. Krauter for Kosmos-Verlag.

Figure 11-1 *Arcella* (0.1 mm)

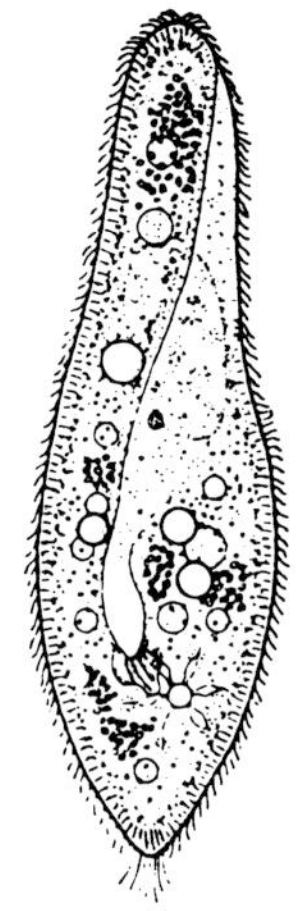

Source: N. Strebel and D. Krauter for Kosmos-Verlag.

Figure 11-2 *Paramecium* (0.2 mm)

Naegleria fowleri, *Euglena* spp., *Tetrahymena* spp., and *Paramecium* spp. (shown in Figure 11-2).

Parasitic protozoa live in or on another organism. The other organism may be either a plant or an animal. This discussion is limited to human enteric (intestinal) protozoans. Some of the enteric protozoa (*Giardia lamblia*, *Cryptosporidium parvum*, *Entamoeba histolytica*, *Cyclospora cayetanensis*, certain microsporidians, and a variety of other protozoa) cause disease within the human small and large intestine, while others do not. Most of these organisms are not considered pathogenic but are commensals, or organisms that live in or on another organism without harming it. *Entamoeba coli* and *Trichomonas hominis* are examples of commensals. Parasitic protozoa, on the other hand, are known to be harmful. Included as representatives of parasitic enteric protozoa are *Entamoeba histolytica*, *Giardia lamblia*, *Cyclospora cayetanensis*, and *Cryptosporidium parvum*.

BIOLOGY AND ECOLOGY

Most enteric protozoa have two stages in their life cycle. The trophozoite is an actively feeding, growing, and reproducing stage. After a while, stimuli in the host's intestinal tract induce most of the enteric protozoans to produce a resistant, dormant transmission form, which is referred to as either a cyst, oocyst, or spore. Whether commensal or parasitic, these protozoa have simple, direct life cycles, and all are transmitted as fecal contaminants of food or water. Person-to-person transmission is also common.

Trophozoites generally do not survive outside their host unless they are in a specialized culture medium. In contrast, cysts, oocysts, and spores may survive for long periods outside the host, especially in cold water. Water temperatures around 25°C or higher and dry conditions reduce the time a cyst, oocyst, or spore can survive outside the host. Only these dormant transmission forms are of concern to the water treatment industry, because they can penetrate the filters. Most troubling for the water treatment industry is the fact that many of these transmission forms are more resistant to chemical disinfection than bacteria and viruses.

Identification

Free-living forms of protozoa are often in sufficient density to be collected in grab samples of either sediment or water close to sediment. Enteric forms, however, are not usually in high enough densities in source waters to allow for grab sampling. Consequently, they must be concentrated for analysis by filtering large volumes of water.

After collection, samples of free-living forms may sometimes be observed directly under the microscope. Free-living amoebae can sometimes be cultured on agar plates containing a massive background of *Escherichia coli* (lawn) that the amoebae use as a food source.

In contrast, the enteric protozoa are difficult to culture in the laboratory. Detection, purification, and cloning of the free-living amoebae are done by isolating organisms from plaques, places where the amoebae have consumed all the bacteria. Parasitic forms, like *Giardia* and *Cryptosporidium* (in Figure 11-3 and in the color section), can be cultured using very exacting aseptic procedures that do not lend themselves to environmental samples. Whatever the organism of interest, the investigator must pay close attention to pH, nutrients present, toxic substances present (such as hydrogen sulfide), temperature, osmotic effects, and other organisms in the sample.

Detection and quantitation of protozoa require skilled microscopic techniques, including epifluorescence, bright-field, phase-contrast, and differential interference microscopy. Moreover, some protozoans are diagnosed based on exacting staining techniques. Very large forms may require observation with a dissecting microscope. Many free-living forms move rapidly and must be slowed down to be seen at all. Substances like Protoslo (Carolina Biological Supply, Burlington, N.C.), methyl cellulose, and polyox resin WSR 301 have been used to increase the viscosity of the medium and slow protozoan movement. Whenever a microscope slide is to be observed for a protracted period, the coverslip must be sealed to the slide with either nail polish, vaspar, or petroleum jelly to prevent evaporation. Air bubbles under a sealed coverslip can provide a limited source of oxygen. For photomicrography of motile living material, a flash attachment for the microscope is obligatory. Identifying *Giardia* cysts and *Cryptosporidium* oocysts requires special equipment and lab personnel specially trained in protozoological identification. Methods are complex and relatively new, and at the time of this writing, there were a limited number of laboratories in the United States providing these analytical services. Because of the US Environmental Protection Agency's Long Term 2 Enhanced Surface Water Treatment Rule, it is anticipated that more laboratories will be offering services for detecting *Giardia* and *Cryptosporidium*.

An abbreviated taxonomic relationship of the protozoans is outlined in Table 11-1.

Giardia lamblia, *Euglena*, *Acanthamoeba*, *Echinamoeba*, *Entamoeba coli*, *E. histolytica*, and *Hartmanella* belong to the phylum Sarcomastigophora, which means amoeboflagellate.

Euglena and *Giardia* are flagellates (subphylum Mastigophora), have vesicular nuclei, and move by hair- or whip-like cylindrical organelles (flagella) that are approximately 0.25 µm wide. Flagellar length and number in this group is variable. Members of this group may also possess chloroplasts, thecal plates, basal bodies, and collars around the flagellum. Some genera exhibit multicellular colonial forms. *Euglena* spp. are examples of free-living flagellates that contain chloroplasts, an identifying characteristic like their single flagellum.

Giardia is an obligate parasite, meaning it must have a host to complete its life cycle. In the environment, *Giardia* has a cyst that is round to oval in shape with dimensions ranging from 8 to 18 µm long by 5 to 15 µm wide. Inside the *G. lamblia* cyst, up to four nuclei, a claw-hammer-shaped median body, and axonemes can be

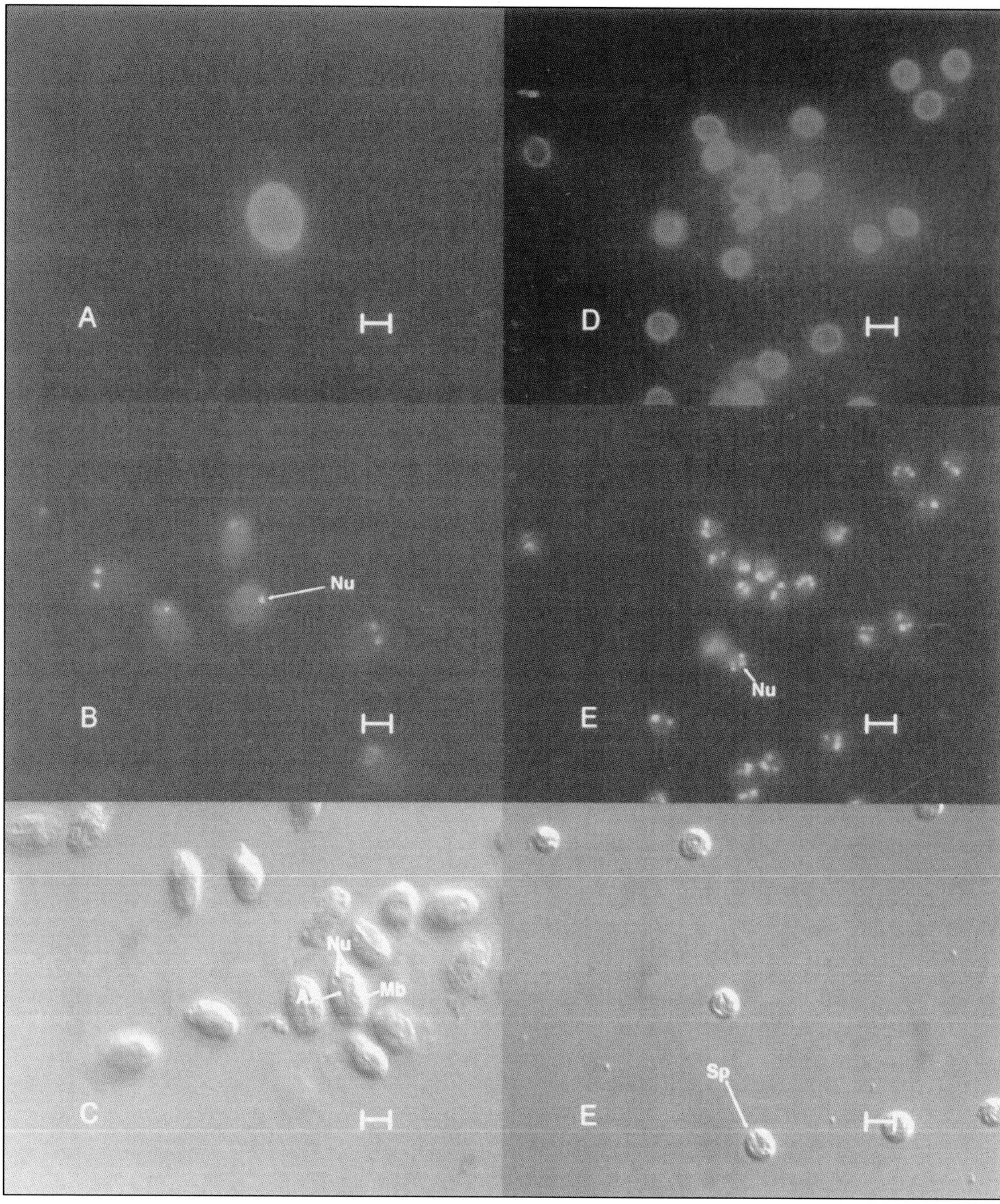

A—Immunofluorescent antibody stained *Giardia* cyst.
B—DAPI stained *Giardia* cysts showing blue fluorescent nuclei (Nu).
C—DIC photomicrograph of *Giardia* cysts. Ax = axonemes, Mb = median body, and Nu = nucleus.
D—Immunofluorescent antibody stained *Cryptosporidium* oocyst.
E—DAPI stained *Cryptosporidium* oocysts showing blue fluorescent nuclei (Nu).
F—DIC photomicrograph of *Cryptosporidium* oocysts. Sp = sporozoite.

Bars = 5 μm

Source: Michael W. Ware, US Environmental Protection Agency.

Figure 11-3 *Giardia* cysts and *Cryptosporidium* oocysts stained by immunofluorescent antibody, 4',6-diamidino-2-phenylindole (DAPI), and Nomarski differential interference contrast (DIC) microscopy

Table 11-1 Taxonomic listing of protozoa potentially associated with potable water

Kingdom: Protista				
Phylum: Sarcomastigophora				
	Subphylum: Mastigophora			
		Class: Phytomastigophora		
			Order: *Euglena* spp.	
		Class: Zoomastigophora		
			Order: Diplomonadida	
			Giardia lamblia	
	Subphylum: Sarcodina			
		Class: Lobosea		
			Order: Amoebida	
			Acanthamoeba castellanii	*Entamoeba histolytica*
			Echinamoeba spp.	*Hartmanella* spp.
			Entamoeba coli	*Naegleria fowleri*
			Order: Arcellinida	
			Arcella spp.	*Quadrulella* spp.
			Cryptodifflugia spp.	
Phylum: Apicomplexa				
		Class: Sporozoea		
			Order: Eucoccidiida	
			Cryptosporidium parvum	*Isospora hominis*
			Cyclospora cayetanensis	*Isospora natalensis*
			Isospora belli	*Toxoplasma gondii*
Phylum: Ciliophora				
		Class: Litostomatea		
			Order: Vestibuliferida	
			Balantidium coli	
		Class: Nassophorea		
			Order: Peniculida	
			Paramecium spp.	
		Class: Oligohymenophorea		
			Order: Hymenostomatida	
			Tetrahymena spp.	
Phylum: Microspora				
		Class: Microsporididea		
		Encephalitozoon		
		Nosema		
		Pleistophora		
		Trachipleistophora		
		Enterocytozoon		

seen (see photographs in color section). *Giardia* cysts are sometimes stained nonselectively with Lugol's iodine, which helps differentiate and facilitate the demonstration of the internal morphological characteristics.

Unfortunately, Lugol's iodine stains everything else in the preparation as well, making it difficult to locate the *Giardia* cysts in water samples. Recently, selective staining of *Giardia* cysts using immunofluorescent antibody (IFA) techniques has become popular. With the use of the IFA staining procedure, demonstration of two internal morphological characteristics by either Hoffman modulation or Nomarski differential interference contrast (DIC) microscopy is desirable. This is because the primary antibody can cross-react with organisms other than *Giardia*, like certain algal cells.

Amoebae belong to the phylum Sarcomastigophora. The subphylum Sarcodina, to which they belong, has members that possess a vesicular nucleus and pseudopods that are retractable cytoplasmic protrusions used both for locomotion and feeding. Thecate amoebae are distinguished from athecate (naked) forms, the other main subgrouping, by a closely fitting envelope or shell secreted by the trophozoite. Many of the thecate amoebae are able to withstand adverse environmental conditions by transforming from a vegetative trophozoite form to a dormant, resistant cyst form. Cysts of certain amoebae have been isolated from the air associated with dust particles. *Hartmanella* spp., *Acanthamoeba castellanii*, *Echinamoeba* spp., and *Naegleria fowleri* are athecate free-living examples that can produce cysts. *Hartmanella* spp., *A. castellanii*, and *Echinamoeba* spp. are known to harbor and amplify *Legionella* spp. *Acanthamoeba* spp. and *N. fowleri* are opportunistic pathogens responsible for primary amoebomeningoencephalitis in humans. Furthermore, *Acanthamoeba* spp. are well-documented causes of corneal keratitis in contact lens wearers.

Entamoeba histolytica and *Entamoeba coli*, as their generic name implies, also are amoebae and belong to the phylum Sarcomastigophora. Both species have cysts that primarily differ from one another on the basis of nuclear morphology. Cysts of *E. histolytica* have four nuclei with centrally located karyosomes (or nucleoli) and are 10–15 µm in diameter. Elongated chromatoid bodies that have bluntly rounded ends are sometimes present in *E. histolytica* cysts. Because *E. coli* cysts can range in size from 10–35 µm in diameter, some of the smaller ones could be confused with *E. histolytica* cysts. However, in the case of *E. coli* cysts, there are 8–16 nuclei with large, eccentrically located karyosomes. Moreover, *E. coli* cysts can have chromatoid bodies that appear splinter-like with pointed ends, but they appear less frequently than the chromatoid bodies in *E. histolytica* cysts.

Using traditional amoeba stains, like Delafield's hematoxylin, besides being laborious, requires a great deal of skill and experience to obtain optimal results. Wet iodine mounts may allow the determination of the nuclear number in *Entamoeba* cysts. In unstained material, phase-contrast microscopy is sometimes helpful in determining the nuclear number, which is the single most important characteristic used in determining species of *Entamoeba*.

Cryptosporidium, *Cyclospora*, *Toxoplasma*, and *Isospora* are genera belonging to the phylum Apicomplexa. This group is totally parasitic and is characterized by (1) an apical complex at the end of certain life-cycle stages; and (2) a sexual phase of the life cycle leading to the production of an oocyst. Besides size and shape of the oocyst, the formation of spores (sporulation) at the time of passage from the host and the number of sporozoites and sporocysts are used to differentiate this group of parasites. Furthermore, the oocysts of this phylum are known to be impervious to traditional stains and resistant to conventional disinfectants.

The *C. parvum* oocyst, like bacteria of the genus *Mycobacterium*, is resistant to most stains with the exception of acid-fast staining. The oocyst of *C. parvum* is round and is 4–6 µm in diameter. Like *Giardia*, IFA staining procedures for the *C. parvum* oocyst are now in the literature, and IFA stains are commercially available. In many instances, either an oocyst wall surface fold or suture line may be detected in IFA-stained material. In most instances, this surface fold is an artifact of staining and is not considered a criterium for detection of this organism. Confirmation of oocysts by DIC microscopy or immunofluorescent 4',6-diamidino-2-phenylindole (DAPI) counter staining to identify one to four sporozoites or 1 to 4 nuclei, respectively, inside the oocyst after detection by IFA is required, because of possible antibody nonspecific staining. When passed by the host, *C. parvum* oocysts are fully sporulated, contain four sporozoites, and the oocysts are infective.

A new coccidian parasite has been isolated from human stool specimens and from surface water. The oocyst of this parasite measures around 8 to 10 μm in diameter. Unlike *Cryptosporidium* oocysts, when passed in the fecal material the *Cyclospora* oocyst must sporulate before becoming infectious. After completing the sporulation process the oocyst contains two ovoid sporocysts measuring approximately 4.0 μm × 6.3 μm, and each sporocyst contains two sporozoites. Like other coccidian parasites, *Cyclospora* oocysts have the acid-fast staining characteristic. Although there are no commercially available antibodies for detecting this parasite, the oocysts of *Cyclospora* have the unique characteristic of autofluorescing when viewed by ultraviolet microscopy.

Three species of *Isospora* are reported as parasitic in humans. They are *I. belli*, *I. hominis*, and *I. natalensis*. Confusion exists regarding the status of *I. belli* and *I. hominis*, as some investigators believe them to be the same organism. There is even a report that suggests *I. hominis* is really a species of *Sarcocystis*. *Isospora natalensis* is rare, occurs only in South Africa, and will not be discussed here.

Unlike *C. parvum* oocysts but like *Cyclospora* oocysts, *I. belli* oocysts are not fully sporulated at the time of evacuation. When sporulation is complete, which requires about 48 hours at room temperature, the oocyst has two sporocysts each containing four crescent-shaped sporozoites. Overall, *I. belli* oocysts are elongated oval structures, measuring 20–30 μm long by 10–19 μm wide with both ends being somewhat narrow. Oocysts of *I. belli* can be detected on the basis of their shape and ability to be acid-fast stained. Like *Cryptosporidium* oocysts and *Giardia* cysts, *I. belli* oocysts can be viewed with phase-contrast and Nomarski DIC microscopy. Although it is unproven, dogs have been suspected of being a reservoir host for *I. belli*.

Isospora hominis oocysts, which are ovoid and 25–33 μm long, are fully sporulated and contain two sporocysts before being evacuated from the host. At the time of passage, the oocyst has broken open and ripe sporocysts containing four sporozoites are passed in the feces. The sporocysts are approximately 14 μm long. Visualization can be by acid-fast staining and bright-field microscopy or by either phase-contrast or Nomarski DIC microscopy with unstained material.

Another enteric coccidian parasite is *Toxoplasma gondii*. This parasite has a complicated life cycle with the sexual phase being in both domestic and wild cats and the asexual phase in the tissues of any mammal. Both sexual and asexual phases also can be completed entirely in the cat. The oocyst, which measures approximately 13 μm × 11 μm, is the result of the sexual cycle in cats and is passed sporadically in their feces. Oocysts of *Toxoplasma* can be detected on the basis of their ability to be acid-fast stained. They also will autofluoresce under ultraviolet light microscopy. The *T. gondii* oocysts can be viewed with phase-contrast and Nomarski DIC microscopy. After sporulation, the oocyst contains two sporocysts each of which contain four sporozoites. There have been two reports of waterborne transmission of *Toxoplasma* in the US.

Ciliates belong to the phylum Ciliophora and are distinguished from other protozoan groups by their unique nuclei. They have both a large macronucleus that regulates cellular metabolism and a small micronucleus that is involved in genetics and sexual recombination. Many ciliates are phagotrophic, which means they ingest nutrients through a mouth or cytosome. These organisms are covered with cilia used in feeding and movement. Cilia are organelles similar in structure to flagella; however, they are generally shorter in length and interconnected through their basal structure. As a result of the interconnection, ciliary movement can be coordinated. *Tetrahymena* spp. and *Paramecium* spp. are examples of free-living ciliates.

Balantidium coli, a parasite of humans, also belongs to the phylum Ciliophora. This is the largest of the enteric parasites with a cyst measuring 40 to 65 μm long. Besides the large size, the cyst is characterized by a thick cyst wall, a micronucleus, and a macronucleus. Young *B. coli* cysts possess cilia that disappear as the cyst ages. While iodine staining of this organism is not recommended because of the density of the cytoplasm, the cyst can be observed with phase-contrast microscopy.

"Microsporidians," a nontaxonomic term, are obligate intracellular parasites belonging to the class Microsporididea of the protozoan phylum Microspora. They are ubiquitous parasites infecting a variety of vertebrate and invertebrate hosts. Insects of commercial significance, such as honeybees and silkworms, are impacted by this group. Moreover, moving up the phylogenetic tree, snails and commercial fish, such as salmon, flounder, and monkfish, are prone to microsporidian diseases. Only with the advent of the AIDS epidemic was this group of pathogens recognized as a cause of human disease. A number of microsporidian genera now have been recovered from humans: *Encephalitozoon*, *Nosema*, *Pleistophora*, *Trachipleistophora*, *Enterocytozoon*, *Septata*, and "*Microsporidium*," a genus for all forms as yet unclassified. Although rare, microsporidian infections are now being reported from immunocompetent people. Presently, classification is done on the basis of small spore size (1.5–5 μm), nuclear configuration, the number of polar tube coils within the spore and developing forms, and the host cell–parasite relationship. The life cycle of these parasites is thought to be direct by either ingestion, inhalation, or inoculation. Two enteric forms isolated from humans, *Enterocytozoon bieneusi* and *Encephalitozoon intestinalis*, may be transmitted by the water route.

SIGNIFICANCE FOR WATER SUPPLIES

Enteric pathogenic protozoa may produce gastrointestinal distress, including diarrhea, flatulence, cramps, anorexia, and weight loss. They can produce a range of symptoms from slight (requiring no medical attention) to acute or chronic (requiring hospitalization).

Presently, both *Giardia* and *Cryptosporidium* are of great concern to the water treatment industry because they are known to have caused a number of waterborne outbreaks of disease. Their control is complicated, because these parasites have several animal hosts other than humans. This increases the numbers of cysts and oocysts that can challenge water treatment plants.

Studies indicate that all surface water supplies should be considered contaminated with both *Giardia* cysts and *Cryptosporidium* oocysts. One of the best barriers in preventing waterborne transmission of enteric protozoa is filtration with pretreatment. Because the smallest of these enteric parasitic protozoa is the *C. parvum* oocyst, properly operated filtration employing a nominal porosity of 1 μm will effectively remove most of these pathogens. The qualifying word "most" is used, because organisms can penetrate even the best run filters. Standard filtration technology used by the water treatment industry will remove 99 percent or more of the cysts in the case of *Giardia*. Studies have shown a little over six logs of *Cryptosporidium* oocysts can be removed using diatomaceous earth filtration. Similarly, when coagulation is properly controlled in conjunction with conventional filtration, total *Cryptosporidium* oocyst removal improves by order of magnitude. In addition, under optimal conventional filtration conditions, *Cryptosporidium* oocyst removal is greater than the total removal of turbidity, particulates, or bacterial spores.

Free-living amoebae that survive the disinfection of water can proliferate in standing-water environments, such as humidifiers, air-conditioning systems, and hot tubs. With increased numbers, *Acanthamoeba* may cause allergic reactions, such as humidifier fever or hypersensitive pneumonitis, when inhaled.

When groundwaters are treated to remove volatile organic compounds through aeration towers, air forced through the towers is not filtered. Some possibility exists that protozoa in groundwater may be seeded with *Legionella* or other bacteria that will then be protected from effective disinfection.

CONTROL STRATEGIES

Because filter penetration is possible, a multiple-barrier approach using watershed protection, pretreatment, filtration, and disinfection for the treatment of surface drinking water is prudent. *Entamoeba histolytica* and *Giardia* cysts are susceptible to both chlorine and combined chlorine, although the concentration-time (*CT*) values for inactivation are high. *CT* values for 99.9 percent inactivation of *G. lamblia* cysts by chlorine have been published in the *Federal Register* (1989). *Giardia lamblia* cysts are relatively susceptible to ozone. Recent information regarding the susceptibility of *C. parvum* oocysts to inactivation by chlorine indicates that the disinfectant concentrations and contact times required are not practical. However, preliminary studies have shown that *C. parvum* oocysts are susceptible to ozone. Whatever chemical disinfectant is employed, factors such as pH, temperature, level of organic particles in the water, and water flow must be considered in order to obtain maximum disinfection efficiency. It is now known that both *Giardia* cysts and *Cryptosporidium* oocysts can be inactivated with ultraviolet light.

Because of the small size of microsporidian spores (1 μm), controlling them by conventional filtration will be quite challenging. On the other hand, because of their large size (10 μm or greater), coccidian parasites other than *Cryptosporidium* should be easily controlled by conventional filtration. Microsporidian spores appear to be susceptible to inactivation with chlorine. Definitive inactivation studies on coccidian parasites like *Cyclospora*, *Toxoplasma*, and *Isospora* have yet to be done.

REFERENCES

Aramini, J.J., C. Stephen, J.P. Dubey, C. Engelstoft, H. Schwantje, and C.S. Ribble. 1999. Potential contamination of drinking water with *Toxoplasma gondii* oocysts. *Epidemiol. Infect.*, 122:305–315.

Ash, L.R., and T.C. Orihel. 1987. *Parasites: A Guide to Laboratory Procedures and Identification.* Chicago, Ill.: American Society of Clinical Pathologists Press.

Beek, W.J., and J.E. Davies. 1976. *Medical Parasitology.* St. Louis, Mo.: C.V. Mosby Co.

Benenson, M.W., E.T. Takafuji, S.M. Lemon, R.L. Greenup, and A.J. Sulzer. 1982. Oocyst-transmitted toxoplasmosis associated with ingestion of contaminated water. *N. Engl. J. Med.*, 307:666–669.

Cavalier-Smith, T. 1993. Kingdom Protozoa and Its 18 Phyla. *Microbiological Reviews*, 57(4):953–994.

Clancy, J.L., W.D. Gollnitz, and Z. Tabib. 1994. Commercial Labs: How Accurate Are They? *Jour. AWWA*, 86:89–97.

Craun, G.F. 1990. Waterborne Giardiasis. In *Giardiasis.* Human Parasitic Diseases. Vol 3. pp. 267–293. E.A. Meyer, ed. Amsterdam, The Netherlands: Elsevier Scientific Publishers.

Despommier, D.D., and J.W. Karapelou. 1987. *Parasite Life Cycles.* New York: Springer-Verlag.

Dugan, N.R., K.R. Fox, J.H. Owens, and R.J. Miltner. 2001. Controlling *Cryptosporidium* Oocysts Using Conventional Treatment. *Jour. AWWA*, 93:64–76.

Faust, E.C., P.F. Russell, and R.C. Jung. 1970. *Craig and Faust's Clinical Parasitology.* 8th ed. Philadelphia, Pa.: Lea and Febiger.

Federal Register. June 29, 1989. (54FR124): 27,532–27,534.

Henke, M., and K. Seidel. 1986. Association Between *Legionella pneumophila* and Amoebae in Water. *Israel J. Med. Sci.*, 22:60–65.

Jakubowski, W. 1984. Detection of *Giardia* Cysts in Drinking Water. State of the Art. In *Giardia and Giardiasis—Biology, Pathogenesis, and Epidemiology.* pp. 263–286. S.I. Erlandsen and E.A. Meyer, eds. New York: Plenum Press.

Jakubowski, W. 1990. The Control of *Giardia* in Water Supplies. In *Giardiasis. Human Parasitic Diseases.* Vol. 3. pp. 335–353. E.A. Meyer, ed. Amsterdam, The Netherlands: Elsevier Scientific Publishers.

Korich, D.G., J.R. Mead, M.S. Madore, N.A. Sinclair, and C.R. Sterling. 1990. Effects of Ozone, Chlorine Dioxide, Chlorine, and Monochloramine on *Cryptosporidium parvum* oocyst viability. *Appl. Environ. Microbiol.*, 56:1423–1428.

Lee, J.J., G.F. Leedale, and P. Bradbury, eds. 2002. *An Illustrated Guide to the Protozoa.* 2nd ed. Lawrence, Kan.: Society of Protozoologists.

Levine, N.D., J.O. Corliss, F.E.G. Cox, G. Deroux, J. Grain, B.M. Honigberg, G.F. Leedale, A.R. Loeblich, J. Lom, D. Lynn, E.G. Merinfield, F.C. Page, G. Poijansky, V. Spragna, L. Vavra, and F.G. Wallace. 1990. A Newly Revised Classification of the Protozoa. *Journal of Protozoology,* 27:37–59.

Ongerth, J.E., and P.E. Hutton. 2001. Testing of Diatomaceous Earth Filtration for Removal of *Cryptosporidium* Oocysts. *Jour. AWWA,* 93:54–63.

Page, F.C. 1976. *An Illustrated Key to Freshwater and Soil Amoebae with Notes on Cultivation and Ecology.* Ambleside, England: Freshwater Biological Association.

Patterson, D.J. 1992. *Free-Living Freshwater Protozoa: A Color Guide.* Boca Raton, Fla.: CRC Press.

Rose, J.B., A. Cifrino, M.S. Madore, C.P. Gerba, C.R. Sterling, and M.J. Arrowood. 1986. Detection of *Cryptosporidium* From Wastewater and Freshwater Environments. *Water Science and Technology,* 18:233–239.

Sauch, J.F. 1985. Use of Immunofluorescence and Phase-contrast Microscopy for Detection and Identification of *Giardia* Cysts in Water Samples. *Appl. Environ. Microbiol.,* 50:1434–1438.

Schaefer, III, F.W. 2001. Can we believe our results? pp. 155–161. In Smith, M. and K.C. Thompson (ed.). *Cryptosporidium*: the Analytical Challenge. Royal Society of Chemistry, Cambridge.

Schaefer, III, F.W. 2002. Detection of protozoan parasites in source and finished drinking waters. pp. 220–233. In *Manual of Environmental Microbiology.* 2nd ed. Hurst, C.J. et al. (eds.) ASM Press, Washington, D.C.

States, S., L. Conley, C. Knezevich, G. Keleti, J. Sykora, R. Wadowsky, and R. Yee. 1988. Free-living Amoebae in Public Water Supplies: Implications for *Legionella*, *Giardia* and *Cryptosporidium*. In *Proceedings of the AWWA Water Quality Technology Conference.* Denver, Colo.: American Water Works Association.

Stetzenbach, L.D., M.J. Arrowood, M.M. Marshall, and C.R. Sterling. 1988. Monoclonal Antibody Based Immunofluorescent Assay for *Giardia* and *Cryptosporidium* Detection in Water. *Water Science and Technology,* 20:193–198.

Streble, H., and D. Krauter. 1982. Das Leben im Wassertropfen. Kosmos, Gesellschaft der Naturfreunde, Franckh'sche Verlagshandlundung, Stuttgart, Germany.

Sykora, J. 1990. Analysis of Water Samples for Protozoans. Methods for the Investigation and Prevention of Waterborne Disease Outbreaks. Gunther Craun, ed. EPA/600 1-990/005a.

Tsieh Sung. 1982. *Pathology and Clinical Features of Parasitic Diseases.* New York: Masson Publishing U.S.A., Inc.

US Environmental Protection Agency. 1992. *Consensus Method for Determining Groundwaters Under the Direct Influence of Surface Water Using Microscopic Particulate Analysis (MPA).* USEPA 910/9-92-029.

US Environmental Protection Agency. 2001. *Method 1623: Cryptosporidium and Giardia in Water by Filtration/IMS/FA.* USEPA 821-R-01-025.

Wickramanayake, G.B., A.J. Rubin, and O.J. Sproul. 1984. Inactivation of *Giardia lamblia* Cysts with Ozone. *Appl. Environ. Microbiol.,* 48:671–672.

Appendix A

AWWA Survey on Nuisance Organisms

This survey was conducted in 1989 by the Organisms in Water Committee of the American Water Works Association to determine the extent of problems experienced by water utilities due to organisms in water. Results compiled from 80 responding utilities follow the sample survey.

AWWA Survey on Nuisance Organisms

The Organisms in Water Committee, Water Quality Division, is updating a manual to help operators identify and treat nuisance organisms. The committee wants to know what organisms have caused problems in your treatment plant or distribution system. To help out, please complete and return the following survey:

Utility Name: ______________________________

Mailing Address: ______________________________

City: ____________________ State: ____________ ZIP: ____________

Person Completing Form: ______________________________

Title: ______________________________

Phone: (________) ______________ FAX: (________) ______________

1. Which organisms have caused problems in your treatment plant and/or distribution system?
   - ☐ No problems experienced.
   - ☐ Bacteria type: ☐ Iron bacteria ☐ Sulfur bacteria ☐ Other ________
   - ☐ Algae type: ☐ Blue–Green ☐ Diatoms ☐ Filamentous ☐ Other ________
   - ☐ Actinomycetes
   - ☐ Protozoa type: ☐ Flagellates ☐ Ciliates ☐ Amoeba ☐ Other ________
   - ☐ Rotifers
   - ☐ Copepods
   - ☐ Hydras
   - ☐ Nematodes type: ☐ Roundworms ☐ Segmented worms
   - ☐ Midge larvae (Bloodworms)
   - ☐ Freshwater jellyfish
   - ☐ Snails
   - ☐ Beetles
   - ☐ Sponges
   - ☐ Moss animals
   - ☐ Water fleas
   - ☐ Clams
   - ☐ Other ______________________________
2. How were you alerted to the problem?
   - ☐ Consumer complaint ☐ Taste-and-odor complaint
   - ☐ Filter-bed biological mass ☐ Open reservoir algal bloom
   - ☐ Flow restrictions in pipe network ☐ Well clogging
   - ☐ Other ______________________________
3. How were the nuisance organisms identified?
   - ☐ Visual sighting ☐ Microscopically ☐ Laboratory cultivation
   - ☐ Other ______________________________
4. What treatment methods were used to eliminate the problem?
   - ☐ Copper sulfate
   - ☐ Increased chlorination
   - ☐ Physical removal type: ☐ Screening ☐ Poly pig ☐ Hydraulic pressure ☐ Other ________
   - ☐ Other treatment ______________________________

   Did the treatment eliminate the problem? ☐ Yes ☐ No
5. Who did you consult for help in identifying and treating the problem organisms?
   - ☐ No outside assistance needed ☐ Another utility
   - ☐ AWWA ☐ University
   - ☐ Private lab ☐ Consultant
   - ☐ Other ______________________________
6. What written references did you use to help you solve the problem?
   - ☐ None ☐ *Standard Methods* ☐ AWWA publication(s) ☐ Other ________

Please return this questionnaire by November 21 to:
P. John Brittan, AWWA, 6666 W. Quincy Ave., Denver, CO 80235

November 1989

1989–1990
Results of AWWA Survey on Nuisance Organisms

1.	Number of surveys returned:		80
2.	Number of utilities that experienced no problems:		7 of 80 (8.75%)
3.	Number of utilities having problems with **iron bacteria:**		21 of 80 (26.25%)

A. How the utility was alerted to the problem?*

1.	Consumer complaint	11	(52.4%)
2.	Taste-and-odor complaint	16	(76.2%)
3.	Clogging of well	8	(38.0%)
4.	Flow restriction in piping network	2	(9.5%)
5.	Laboratory testing	2	(9.5%)
6.	Visual	1	(4.8%)

B. How were the nuisance organisms identified?*

1.	No answer	1	(4.8%)
2.	Microscopically	10	(47.6%)
3.	Laboratory cultivation	9	(42.8%)
4.	Visual sighting	5	(23.8%)
5.	Outside laboratory	2	(9.5%)

C. Treatment methods used and if they were successful*

			Successful			
	Treatment	Quantity	Yes	No	Somewhat	No Answer
1.	Increase chlorination	12	3	3	3	3
2.	Flushing	4	1	1	2	0
3.	Shock chlorination	3	0	0	1	2
4.	Poly pig	2	0	0	2	0
5.	Greensand filter	2	0	1	0	1
6.	Upgrade well	1	0	0	0	1
7.	No treatment	2	0	0	0	2
8.	Aqua mag	2	0	1	1	0

D. Who did the utilities consult to help identify and solve the problem?*

1.	Another utility	4
2.	State or USEPA laboratory	4
3.	Consultant	5
4.	Private lab	2
5.	University	3
6.	No outside assistance needed	8

*More than one answer given by many utilities.

E. What written references were used to solve the problem?*

1.	*Standard Methods*	9
2.	AWWA publications	5
3.	Other publications (unnamed)	4
4.	None	10

4. Number of utilities having problems with **sulfur bacteria**: 7 of 80 (8.75%)

A. How the utility was alerted to the problem?*

1.	Consumer complaint	4	(57.2%)
2.	Taste-and-odor complaint	6	(85.7%)
3.	Well clogging	3	(42.9%)

B. How were the nuisance organisms identified?*

1.	Microscopically	4	(57.2%)
2.	Laboratory cultivation	3	(42.9%)
3.	Visual sighting	3	(42.9%)
4.	Smell	1	(14.2%)
5.	No answer	1	(14.2%)

C. Treatment methods used and if they were successful*

			Successful			
	Treatment	Quantity	Yes	No	Somewhat	No Answer
1.	Increase chlorination	6	3	1	1	1
2.	Shock chlorination	3	2	0	0	1
3.	Aeration	1	1	0	0	0
4.	Flushing	1	1	0	0	0
5.	Poly pig	1	0	0	1	0

D. Who did the utilities consult to help identify the problem?*

1.	Another utility	1
2.	State or USEPA laboratory	2
3.	Consultant	2
4.	No outside assistance needed	2

E. What written references were used to solve the problem?*

1.	*Standard Methods*	2
2.	AWWA publications	0
3.	None	4
4.	Handbook of Chlorination	1
5.	Bergley's Sulfate-Reducing Bacteria	1

*More than one answer given by many utilities.

5. Other types of bacteria identified by utilities which have caused problems?*

1.	Nitrifying	2
2.	Nitrobacter	1
3.	Nitrosomonas	1
4.	Enterobacter	1
5.	Coliform	3
6.	SPC	1
7.	*E. Cloacae*	1
8.	Flavobacteria	1

6. Number of utilities having problems with **Algae**: 48 of 80 (60%)

A. Types of bacteria identified

1.	Blue–green	35	(73%)
2.	Diatoms	22	(46%)
3.	Filamentous	20	(42%)
4.	Synura	1	(2%)
5.	Green	3	(6%)
6.	Dinoflagellates	2	(4%)

B. How the utility was alerted to the problem?*

1.	Consumer complaint	16	(33%)
2.	Taste-and-odor complaint	31	(65%)
3.	Operational monitoring	8	(17%)
4.	Open reservoir algal bloom	25	(52%)
5.	Routing microscopic examination	4	(8%)
6.	Filter-bed biological mass	10	(21%)
7.	Short filter runs	3	(6%)
8.	Clogging of POU filters	1	(2%)

C. How were the nuisance organisms identified?*

1.	Visual sighting	22	(46%)
2.	Microscopically	41	(85%)

D. Treatment methods used and if they were successful*

			Successful			
	Treatment	Quantity	Yes	No	Somewhat	No Answer
1.	Increase chlorination (↑ Cl_2)	6	2	3	1	0
2.	Potassium permanganate ($KMnO_4$)	2	2	0	0	0
3.	Powdered activated carbon (PAC)	5	4	0	1	0
4.	Copper sulfate ($CuSO_4$)	14	7	0	7	0
5.	↑ Cl_2/PAC	2	0	0	2	0
6.	↑ Cl_2/PAC/$CuSO_4$	2	2	0	0	0
7.	↑ Cl_2/$CuSO_4$	14	7	3	4	0
8.	↑ Cl_2/$KMnO_4$/PAC	2	2	0	0	0
9.	Screening	2	0	0	2	0
10.	Treatment change	2	0	1	1	0
11.	Hydraulic pressure	2	0	0	2	0
12.	Flushing	4	2	0	2	0

*More than one answer given by many utilities.

E. Who did the utilities consult to help identify and solve the problem?*

1.	AWWA	5	(10%)
2.	Private lab	2	(49%)
3.	State	6	(13%)
4.	University	11	(23%)
5.	Another utility	12	(25%)
6.	Consultant	4	(8%)
7.	Seminars	2	(4%)
8.	Chemical manufacturer	2	(4%)
9.	USEPA	1	(2%)
10.	No assistance required	22	(46%)

F. What written references were used to solve the problem?*

1.	None	10
2.	*Standard Methods*	29
3.	AWWA publications	28
4.	Other unnamed materials	1
5.	Illinois seminar materials	1
6.	Phelps Dodge manual	1
7.	Book by Weber	1
8.	Algae identification manual	1
9.	Old USEPA algae publication	1
10.	Book by Smith	1
11.	Photos taken by phase-contrast microscope	1

7. Number of utilities having problems with **Actinomycetes**: 21 of 80 (26.25%)

A. How the utility was alerted to the problem*

1.	Consumer complaint	10	(48%)
2.	Filter-bed biological mass	7	(33%)
3.	Taste-and-odor complaints	16	(76%)
4.	Algal bloom	10	(48%)

B. How were the nuisance organisms identified?*

1.	Microscopically	15	(71%)
2.	Laboratory cultivation	9	(43%)
3.	Presumed	4	(19%)
4.	Chemical analysis	1	(5%)

C. Treatment methods used and if they were successful*

	Treatment	Quantity	Successful			
			Yes	No	Somewhat	No Answer
1.	PAC	2	1	0	1	0
2.	GAC	1	1	0	0	0
3.	$KMnO_4$/PAC	3	1	1	1	0
4.	↑ Cl_2	8	0	5	3	0
5.	↑ Cl_2/PAC	1	0	0	1	0
6.	↑ Cl_2/$CuSO_4$	4	2	0	2	0
7.	$KMnO_4$/$CuSO_4$	1	0	0	1	0

*More than one answer given by many utilities.

8. Number of utilities having problems with **Midge larvae (Bloodworms)**: 13 of 80 (16.25%)

A. Other types of larvae identified

1. Mosquito
2. Black fly
3. Chaoborus

B. How the utility was alerted to the problem*

1.	Filter-bed biological mass	2	(16%)
2.	Consumer complaint	10	(77%)
3.	Routine monitoring	1	(8%)

C. Treatment methods used and if they were successful*

			Successful			
	Treatment	Quantity	Yes	No	Somewhat	No Answer
1.	↑ Cl_2	10	5	3	2	0
2.	↑ Filter backwashing	2	2	0	0	0
3.	Flushing	3	0	1	2	0

9. Number of utilities having problems with **Protozoa**: 9 of 80 (11.25%)

A. Types of protozoa identified

1.	Flagellates	7	(78%)
2.	Ciliates	4	(44%)
3.	Amoeba	2	(22%)

NOTE: Other information (treatment, identification, etc.) overlaps with other organisms.

10. Other organisms identified by utilities as ones that cause problems*

A.	Rotifers	4	(5%)
B.	Copepods	5	(6.25%)
C.	Hydras	1	(1.25%)
D.	Nematodes	4	(5%)
	1. Roundworms	2	(2.5%)
	2. Segmented worms	2	(2.5%)
E.	Freshwater jellyfish	0	(0%)
F.	Snails	1	(1.25%)
G.	Zebra mussels	1	(1.25%)
H.	Fungus	1	(1.25%)
I.	Water fleas	7	(8.75%)
J.	Beetles	1	(1.25%)
K.	Sponges (Byrzoan)	2	(2.5%)
L.	Clams	3	(3.75%)

*More than one answer given by many utilities.

This page intentionally blank.

Appendix B

Troubleshooting Guide for Problem Organisms

The following chart (Table B-1) condenses material presented in this manual for easy reference. For additional information on the organisms listed, refer to the specific chapter for each organism. It is important to note that only the organisms discussed in the text appear in the chart and there are often more causes for a problem than these organisms. For example, a taste or odor problem may be caused by water from a different source, watershed changes, cross connections, old piping, or disinfecting after repairs in the distribution system, to name only a few possibilities.

Table B-1 Troubleshooting guide for problem organisms

Problem	Symptom	Probable Solution
Actinomycetes (Chapter 1)	Earthy/musty/moldy tastes and odors.	Ozone, hydrogen peroxide, and ozone.
		In source water: alternate source, vary intake depth, activated carbon, copper sulfate.
		In treatment plant: minimize sludge depth, regular basin cleaning.
		In distribution system: activated carbon, flushing, superchlorination, pigging, and superchlorination.
Algae (Chapter 10)	Clogged filters.	In source water: copper sulfate, oxygenation, nutrient control.
		In treatment plant: activated carbon, ozone, and chlorination (not for some types of algae).
	Reservoir oxygen depletion/fish kills.	Copper sulfate, oxygenation, nutrient control.
	Tastes and odors. See text for specific algae and associated taste/odor.	See algae, blocked filters.
	Toxicity.	Activated carbon.
Bloodworms (midges, chironomids) nonparthenogenetic (for parthenogenetic see text) (Chapter 6)	Visible organisms.	Increase chlorination and backwash, flushing.
Crustacea (Chapter 7)	Clogged filters.	Flushing. Increase filtration efficiency. Disinfectants targeted to specific organisms.
	Color.	See crustacea, clogged filters.
	Interference with disinfection.	See crustacea, clogged filters.
	Tastes and odors.	Ozone, activated carbon.
	Turbidity increase.	See crustacea, clogged filters.
	Visible organisms.	See crustacea, clogged filters.
Iron bacteria (Chapter 2)	Color.	Prevent construction flaws. Regular chlorination. Install filtration.
	Corrosion.	See iron bacteria, color.
		In wells: superchlorination, physical agitation, acid dissolution, surfactant dispersion, others (see text).
		In filters: backwashing, superchlorination, increased prechlorination, greensand filtration.
		In distribution system: superchlorination, install loops in system, physical removal of bacteria, corrosion chemical treatment, filtration, chlorination.

Table continued next page

Table B-1 Troubleshooting guide for problem organisms (continued)

Problem	Symptom	Probable Solution
Iron bacteria (Chapter 2)	Disinfectant demand increase.	See iron bacteria, color.
	Frothing.	See iron bacteria, color.
	Reduced capacity.	See iron bacteria, corrosion.
	Tastes and odors.	See iron bacteria, color.
	Turbidity increase.	See iron bacteria, color. Turbidity monitoring.
Nematodes (Chapter 5)	Interference with disinfection.	Increase chlorination and backwash, flushing. Optimize treatment processes, especially settling.
	Visible organisms.	See nematodes, interference with disinfection.
Nitrifying bacteria (Chapter 4)	Disinfectant demand increase.	Flushing. Superchlorinating reservoirs and storage tanks. Increasing Cl_2:N weight ratio. Decreasing detention time in reservoirs and distribution. Chlorine disinfection.
Protozoa (Chapter 11)	Enteric disease.	Disinfection specific to organism, multiple-barrier approach to protect water quality, and optimizing plant performance.
Rotifers (Chapter 8)	Visible organisms.	Deactivate before treatment train. Optimize coagulation, flocculation, and filtration. Cover finished water reservoirs.
Sulfur/hydrogen sulfide bacteria (Chapter 3)	Color.	Disinfection, acidification, and cleaning. Increase dissolved oxygen.
	Corrosion.	Routine use of disinfectant and penetrant. Minimize pump inactivity.
	Reduced capacity.	See sulfur bacteria, color.
	Rotten-egg odor.	Aeration, chlorination, sulfur dioxide.
Zebra mussels (Chapter 9)	Reduced capacity.	Chlorination, ozonation. Antifouling coatings. Physical scraping, pigging. High-pressure wash. All methods require removal of dead and dislodged zebra mussels.
		At intake: sand infiltration beds, increase temperature, strainers, screens, or filters.
		In treatment plant: oxygen deprivation.
	Tastes and odors.	Remove decaying organisms, disinfection.

This page intentionally blank.

Appendix C

Suggestions for Optimizing Conventional Water Treatment

The following table (Table C-1) is intended to assist conventional water treatment system operators in troubleshooting problems occurring throughout the system, from the supply source through the distribution system. The table is derived from an article by W.E. Bellamy, J.L. Cleasby, G.S. Logsdon, and M.J. Allen, which appeared in the December 1993 issue of *Journal AWWA*.

Table C-1 Suggestions for optimizing conventional water treatment systems

Water supply source	Select and protect the best available source.
	Evaluate source contamination from agriculture, wastewater, roads, railroads, industry, homes, pipelines, etc., in drainage area.
	Consider alternative sources: groundwater, second plant, neighboring water system.
	Change location of intake and/or depth of drawoff or install multilevel intake.
	Monitor microbiological parameters regularly.
Rapid mix/coagulation	Check mixing energy, time, dispersion of coagulant.
	Check design flow rate and do not exceed.
	Check condition of equipment, even flow splitter and baffling before flocculation.
	Optimize coagulant dose (jar test, zeta potential).
	Optimize coagulant conditions (pH, alkalinity).
	Consider coagulant aid and alternate coagulants.
	Install coagulant control device (streaming current detector, pilot filter, zeta potential).
	Add flow-paced control for coagulants.
Flocculation	Check mixing energy and time.
	Check conditions of equipment and baffling (between chambers and before settling).
	Consider flocculant aid polymer.
	Check design flow rate to prevent floc shearing.
Sedimentation	Check design parameters and compare to current plant operations: flow rate, detention time, weir loading (arrangement, drawoff, and flow splitting).
	Check condition of sludge removal equipment and weirs.
	Check for short circuiting.
	Check and improve baffle between flocculation and sedimentation basins.
	Check sludge drawoff procedures.
	Maintain low sludge levels; waste or blowdown sludge when necessary.
	Do not disrupt sludge blanket (changes in flow, temperature, wind, sludge blowdown rate).
Filtration	Assess filter backwash and wastewater recycle practices; treat backwash effluent; evaluate recycle quality for return to supply.
	Avoid sudden changes in filter rate (do not exceed design flow rate, minimize plant flow rate changes, adjust filter control valve to operate properly and observe valve operation, bring another filter on line when one is being backwashed).
	Measure filter flow rate and compare with specifications.
	Establish as low a turbidity goal as possible (0.1 ntu or less).
	Install continuously monitoring turbidimeters.
	Consider particle counters to monitor particle removal.
	Evaluate filter media (size, uniformity coefficient, shape, gravel mounding, mud balls, depth, surface cracks).
	Evaluate underdrains (type, condition, backwash distribution, plugging).
	Check for uniform flow during backwash and gravel upset.
	Evaluate surface wash and/or air-water backwash equipment.
	Review backwash procedure (flow rates, sequence, duration, criteria for ending backwash, operational changes for water temperature).
	Allow media to settle after backwash before bringing filter on line.

Table continued next page

Table C-1 Suggestions for optimizing conventional water treatment systems (continued)

Filtration (continued)	Consider filter-to-waste capability, if not already installed. Bring filters on line slowly. Do not bring several filters on line at the same time. Do not bring a filter back into operation without backwashing. Consider filter aids and adding coagulant to backwash water. Establish criteria for initiating filter backwash (time, head loss, turbidity, particle counts). Monitor microbial parameters. Disinfect filters. Treat each filter as a separate unit process.
Disinfection	Evaluate disinfection effectiveness ($C{\times}T$). Consider ozone as a second barrier for *Cryptosporidium* inactivation.
Distribution	Flush dead ends and mains. Routinely drain, flush, inspect, and disinfect reservoirs, basins, and elevated tanks. Establish hydrant and valve maintenance programs. Organize and train staff on emergency plan for breaks. Produce stable water to discourage corrosion. Maintain coupon testing to monitor corrosion. Sample distribution system to monitor corrosion.
Miscellaneous	Consider pilot plant studies to facilitate optimization. Consider temporarily downrating the plant to improve performance. Establish standard operating procedures for all unit processes. Consider establishing a department that is responsible for water quality from source to tap. Establish maintenance management system for all facilities.

Source: W.E. Bellamy, J.L. Cleasby, G.S. Logsdon, M.J. Allen. 1993. Assessing Treatment Plant Performance. Jour. AWWA, *85:12:34–38.*

This page intentionally blank.

Appendix D

Actinomycete Culture Agars

Regular sampling and examination for actinomycetes provide baseline data for a utility experiencing problems with these organisms. With these data, system personnel may predict problems as they note increases in actinomycete populations and make operational changes before actinomycetes become a problem. Although other media will grow actinomycetes, the five media types that follow are considered the most convenient to make and use.

ACTINOMYCETE ISOLATION AGAR*

Reagents

Sodium caseinate	2.0 g
Asparagine	0.1 g
Sodium propionate	4.0 g
Dipotassium phosphate	0.5 g
Magnesium sulfate	0.1 g
Ferrous sulfate	0.001 g
Agar	15.0 g
Final pH: 8.1	

Procedure

Suspend the ingredients in 1 L of distilled water and heat until boiling to dissolve completely. Add 5 g of glycerol and autoclave for 15 min at 15 psi and 121°C. Allow to cool to 55 to 60°C and pour into 100-mm Petri dishes. Inoculate the surface of the cooled plate, by the spread plate technique, with the appropriate amount of sample. Incubate at 30°C for 40 to 72 hr.

*Available from Difco Laboratories, Detroit, Mich.

EGG ALBUMIN AGAR

Reagents

Glucose	1.0 g
Soluble egg albumin	0.25 g
Dipotassium phosphate	0.5 g
Magnesium sulfate	0.2 g
Ferrous sulfate	0.01 g
Agar	15.0 g
Final pH: 7.1	

Procedure

Suspend the ingredients in 1 L of distilled water and heat until boiling to dissolve completely. Autoclave for 15 min at 15 psi and 121°C. Allow to cool to 55 to 60°C and pour into 100-mm Petri dishes. Inoculate the surface of the cooled plate, by the spread plate technique, with the appropriate amount of sample. Incubate at room temperature for 5 to 14 days. Plates should be inspected on a routine basis to observe the appearance of the first colonies.

SODIUM CASEINATE AGAR

Reagents

Glucose	1.0 g
Sodium caseinate	2.0 g
Dipotassium phosphate	0.2 g
Magnesium sulfate	0.2 g
Ferrous sulfate	0.01 g
Agar	15.0 g
Final pH: 7.3	

Procedure

Suspend the ingredients in 1 L of distilled water and heat until boiling to dissolve completely. Autoclave for 15 min at 15 psi and 121°C. Allow to cool to 55 to 60°C and pour into 100-mm Petri dishes. Inoculate the surface of the cooled plate, by the spread plate technique, with the appropriate amount of sample. Incubate at room temperature for 5 to 14 days. Plates should be inspected on a routine basis to observe the appearance of the first colonies.

STARCH AGAR

Reagents

Soluble starch	10.0 g
Dipotassium phosphate	1.0 g
Magnesium sulfate	1.0 g
Ammonium sulfate	2.0 g
Calcium carbonate	2.0 g
Trace elements*	1.0 mL
Agar	20.0 g

*Trace elements solution	
Ferrous sulfate	0.1 g
Manganous chloride	0.1 g
Zinc sulfate	0.1 g
Distilled water	1,000 mL

Procedure

Suspend the ingredients in 1 L of distilled water and heat until boiling to dissolve completely. Autoclave for 15 min at 15 psi and 121°C. Allow to cool to 55 to 60°C and pour into 100-mm Petri dishes. Inoculate the surface of the cooled plate, by the spread plate technique, with the appropriate amount of sample. Incubate at room temperature for 5 to 14 days. Plates should be inspected on a routine basis to observe the appearance of the first colonies.

M1B2 AGAR

Reagents

Sodium citrate	10.0 g
Glucose or dextrose	10.0 g
Ferrous sulfate	0.01 g
Magnesium sulfate	0.05 g
Calcium chloride	0.1 g
Ammonium nitrate	6.0 g
Disodium phosphate	2.0 g
Agar	15.0 g

Procedure

Suspend the ingredients in 1 L of distilled water and heat until boiling to dissolve completely. Autoclave for 15 min at 15 psi and 121°C. Allow to cool to 55 to 60°C and pour into 100-mm Petri dishes. Inoculate the surface of the cooled plate, by the spread plate technique, with the appropriate amount of sample. Incubate at room temperature for 5 to 14 days. Plates should be inspected on a routine basis to observe the appearance of the first colonies.

This page intentionally blank.

Appendix E

Iron Bacteria Presence/Absence and Quantity Methods

The following methods for determining the presence or absence and quantity of iron bacteria are based on Section 9240 of *Standard Methods for the Examination of Water and Wastewater.*

The American Society of Testing and Materials* (ASTM) publishes a similar method entitled D932-85 *Standard Test Method for Iron Bacteria in Water and Water-Formed Deposits.*

A. Methods to Determine Absence or Presence of Iron Bacteria

1. If there is no visible material, the water sample may be centrifuged or settled overnight to concentrate the cells prior to examination. Place a drop of concentrate on a slide, add a coverslip, and examine the sample at 1,000× (400× is often sufficient magnification). If conventional light microscopy is used, stain preparations with India ink or lactophenol blue. Many iron bacteria structures and deposits are clearly visible without stain. Phase-contrast microscopy permits easier examination of unstained material.

   The water sample may also be filtered through a 0.45-µm pore size membrane filter to concentrate the cells. Dry the filter, clear it with immersion oil, and examine at 1,000×.

   Glass microscopic slides also can be placed directly in the well or water system in protective collectors for direct microscopic examination. Typical exposure time is for one week, although time is site specific.

   Heavy deposits of iron outside the cells can obscure the view of iron bacteria. To overcome this, dissolve iron deposits in HCl, oxalic, or citric acid. Citric acid does not cause lysis of cell material but is less effective in Fe removal. Add a coverslip, then place a few drops of 1*N* HCl or 0.5–1 percent citric acid on one side of the coverslip. Draw acid under the coverslip by touching a piece of blotting paper to the opposite side of the coverslip. Acid will dissolve the iron deposits, and the bacteria can be observed. Deposits can be identified as iron by adding

*American Society for Testing and Materials, 1916 Race St., Philadelphia, PA 19103.

a few drops of potassium ferrocyanide to a sample on a slide. A blue precipitate forms when iron in or around the cells is dissolved.

2. Examine slime, floc, or other visible materials directly by placing a portion of the material under a coverslip. The sample should be treated as described above in regard to staining for examination by light microscopy, dissolving deposits with acid, and reacting with potassium ferrocyanide to determine the presence of iron.

B. Methods to Determine Quantity of Iron Bacteria

1. Exact quantitation is difficult due to the particulate nature of the material, but a semiquantitative evaluation can be made. This "quantification" (e.g., ASTM method D932-85) is limited only to the sample and does not represent what is happening throughout the system. Samples are usually fragments shed from biofilms on the surfaces of rock, pumps, and pipe and are not indicative of the entire system.

   Presence of heterotrophic Fe bacteria may be determined using Fe-amended HPC media. Brown or orange colonies can be counted as in any plate-count procedure.

2. Another semiquantitative method has been proposed using commercial dehydrated media. Days until reaction, such as cloudiness, can indicate bacterial concentrations in the medium.

Appendix F

Color Section

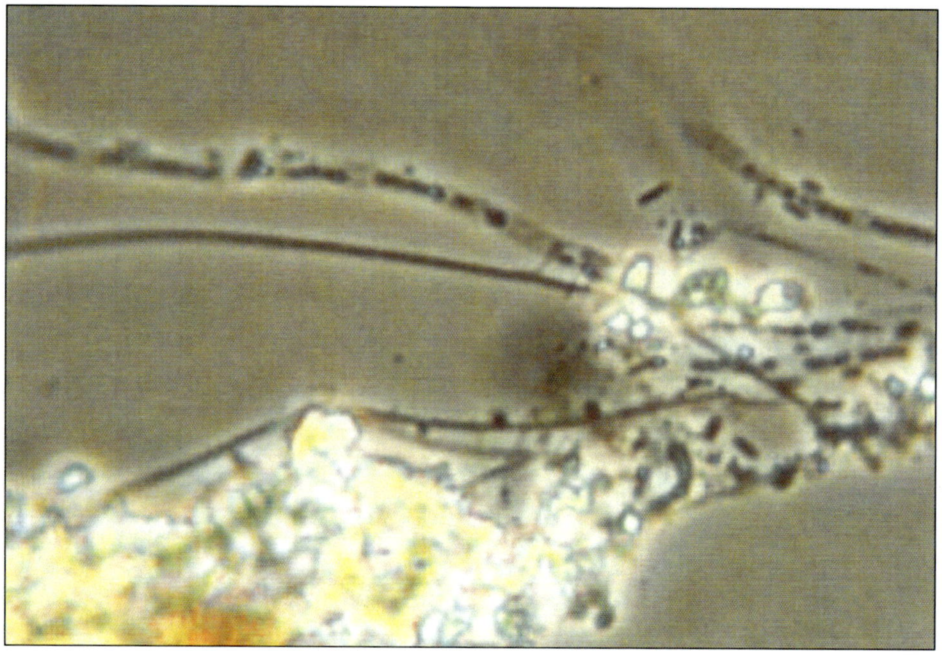

Source: M. Richard, PhD, RBD Inc., Ft. Collins, Colo.

Figure F-1 Iron bacteria. *Leptothrix* (note sheath), 1,000×

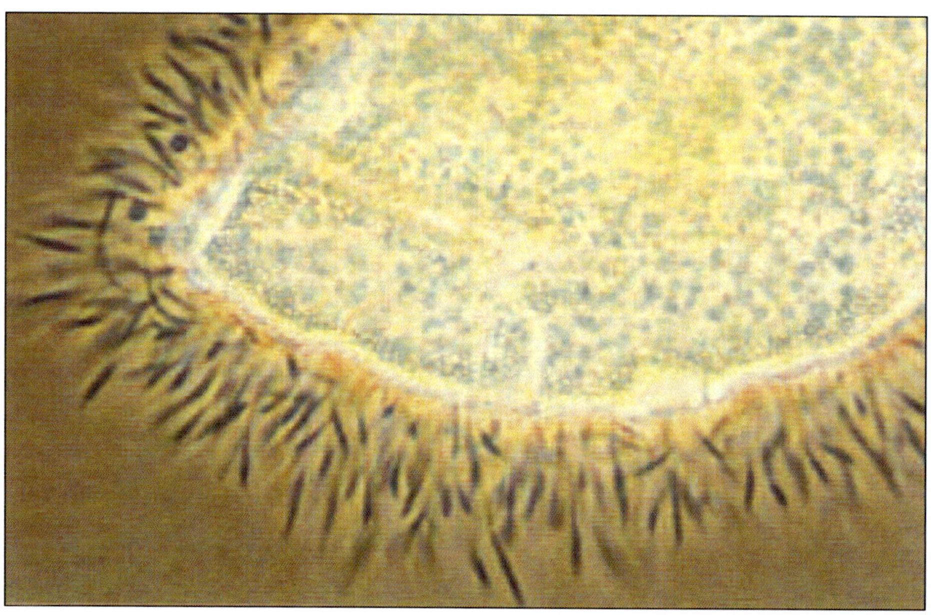

Source: M. Richard, PhD, RBD Inc., Ft. Collins, Colo.

Figure F-2 Iron bacteria. *Caulobacter* (large "colony"), 1,000×

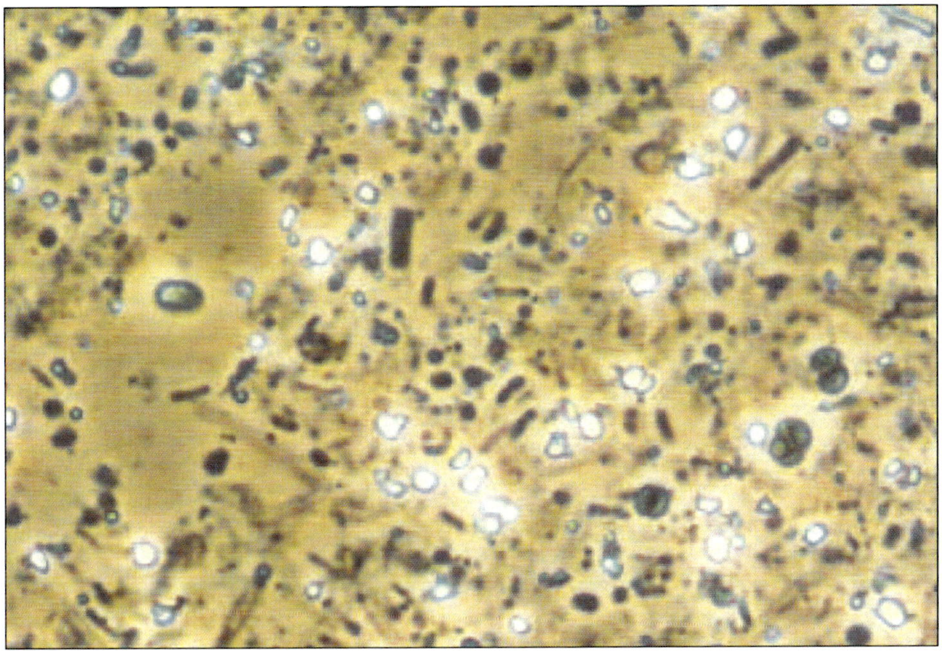

Source: *M. Richard, PhD, RBD Inc., Ft. Collins, Colo.*

Figure F-3 Iron bacteria. *Thiobacillus ferro oxidans* (note precipitated iron), 1,000×

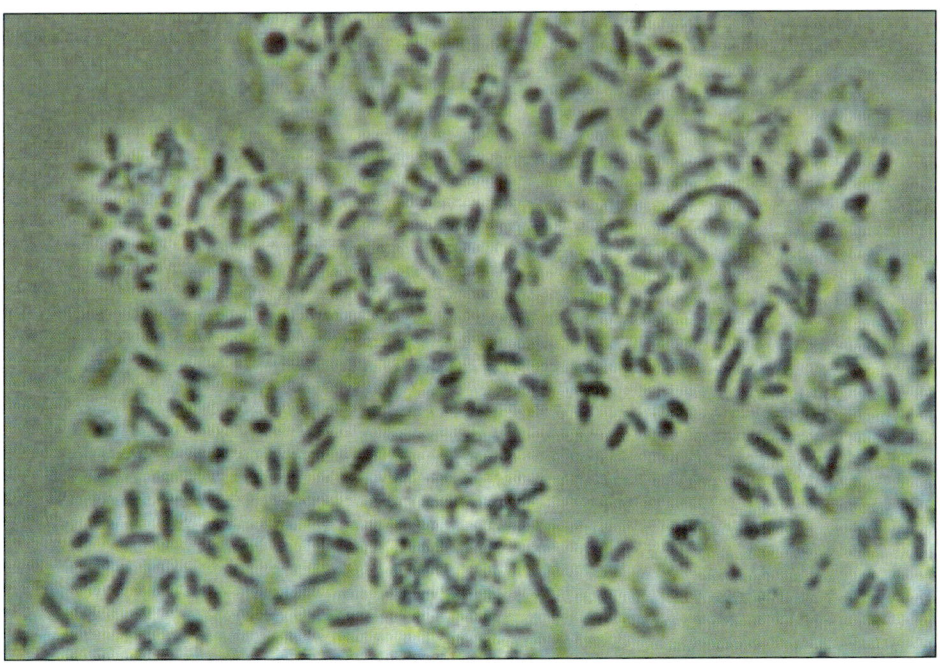

Source: *M. Richard, PhD, RBD Inc., Ft. Collins, Colo.*

Figure F-4 Iron bacteria. *Pseudomonas,* 1,000×

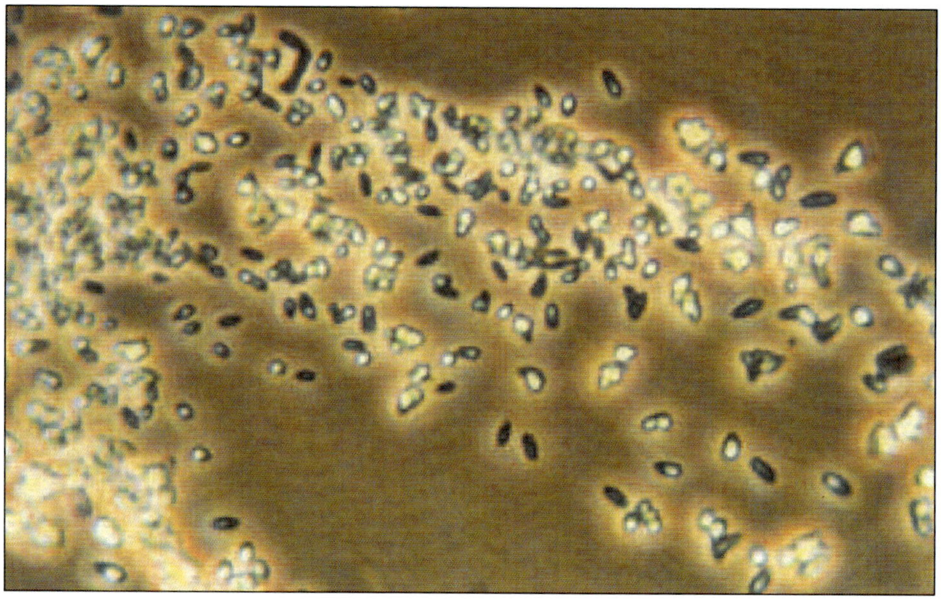

Source: M. Richard, PhD, RBD Inc., Ft. Collins, Colo.

Figure F-5 Sulfur bacteria. *Thiobacillus* (note precipitated sulfur), 1,000×

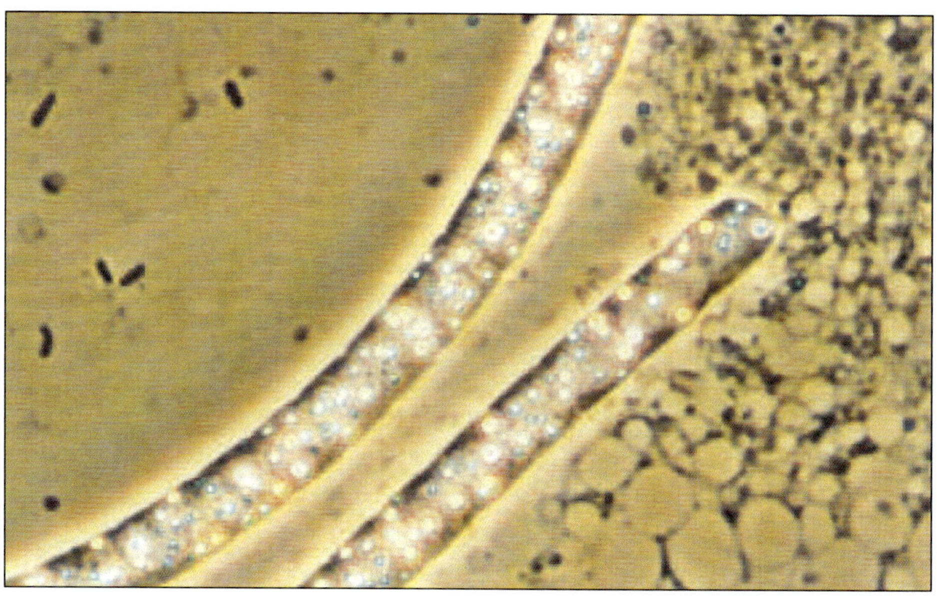

Source: M. Richard, PhD, RBD Inc., Ft. Collins, Colo.

Figure F-6 Sulfur bacteria. *Beggiatoa* (note sulfur granules), 1,000×

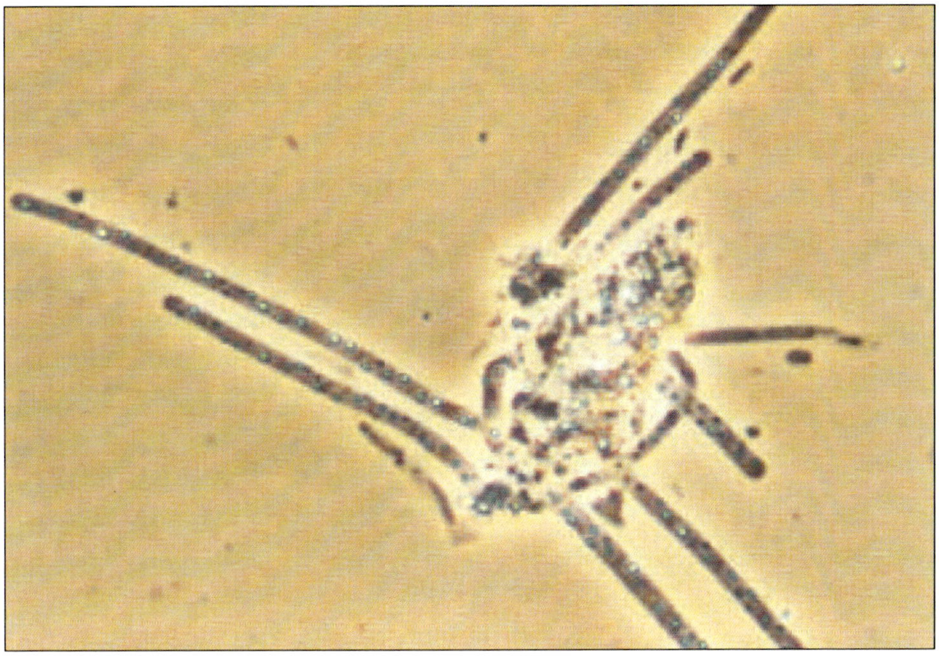

Source: M. Richard, PhD, RBD Inc., Ft. Collins, Colo.

Figure F-7 Sulfur bacteria. *Thiothrix* (note sulfur granules), 1,000×

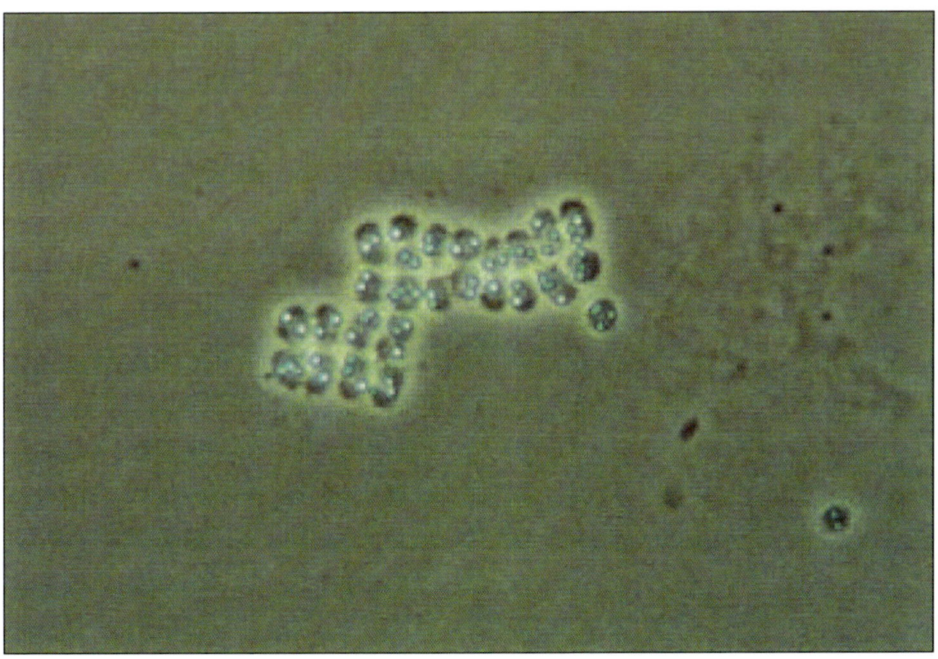

Source: M. Richard, PhD, RBD Inc., Ft. Collins, Colo.

Figure F-8 Sulfur bacteria. *Thiopedia* (note sulfur granules), 1,000×

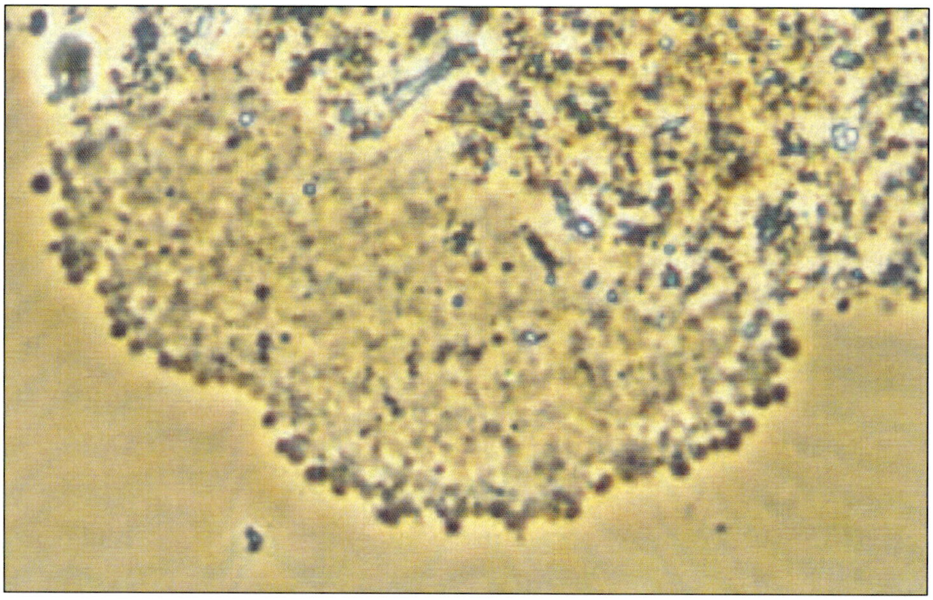

Source: M. Richard, PhD, RBD Inc., Ft. Collins, Colo.

Figure F-9 Nitrifying bacteria. *Nitrosomonas* (cocci), 1,000×

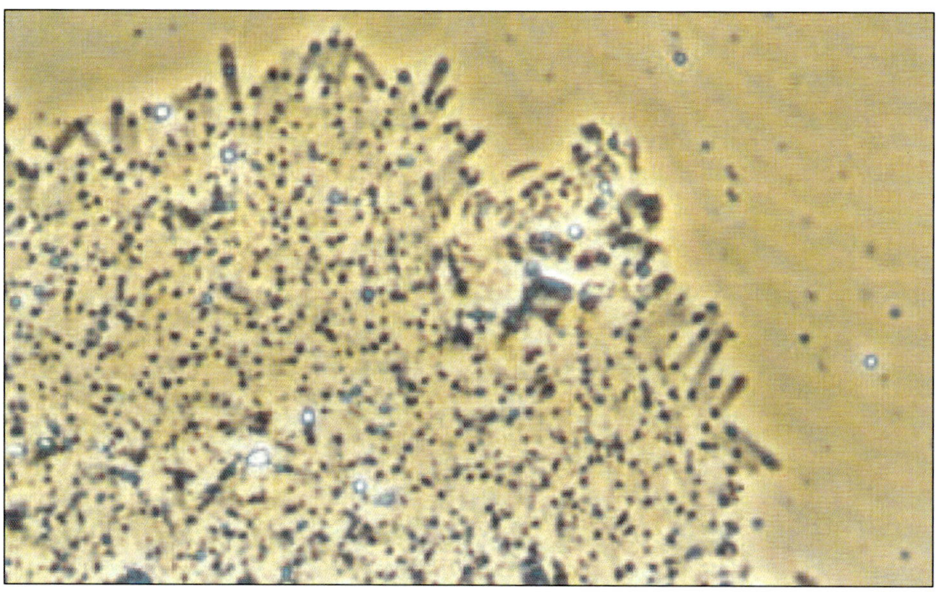

Source: M. Richard, PhD, RBD Inc., Ft. Collins, Colo.

Figure F-10 Nitrifying bacteria. *Nitrobacter* (rods), 1,000×

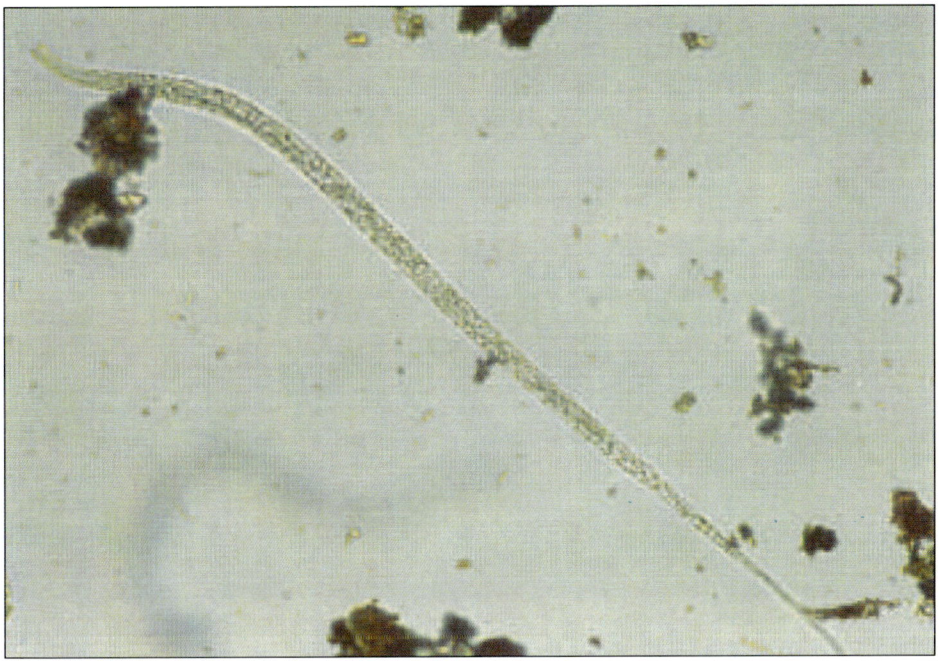

Source: K. Hancock, CHDiagnostic & Consulting Service Inc.

Figure F-11 Nematode larva, 60×

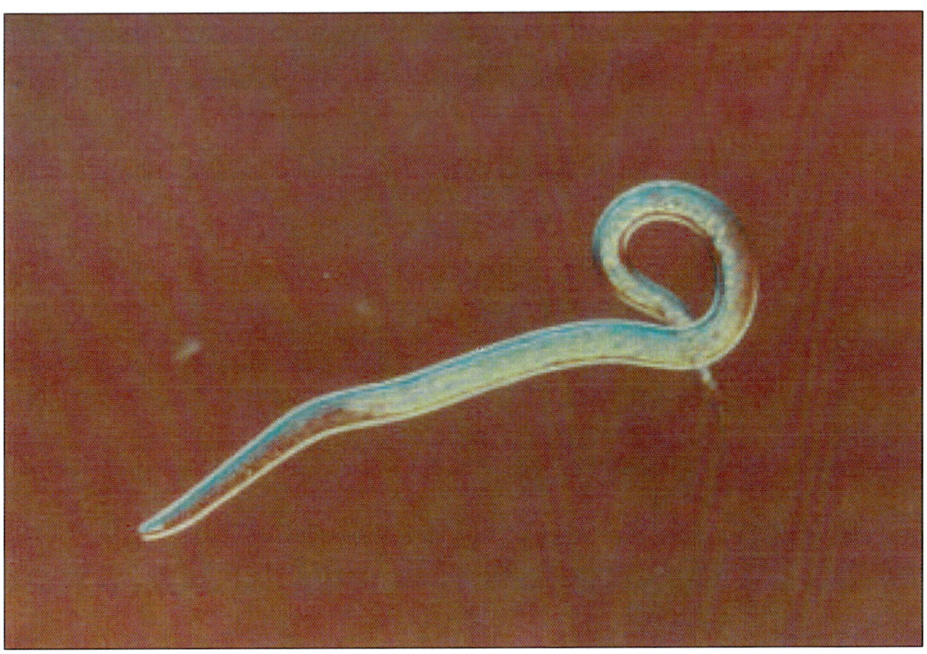

Source: R.C. Lorenz, City of Westerville, Ohio.

Figure F-12 Nematode, 250×

Source: Drs. J.H.M. van Lieverloo, Kiwa N.V. Research and Consultancy, Nieuwegein, The Netherlands.

Figure F-13 Chironomid. Head of a pupa, dorsal view, 1:7

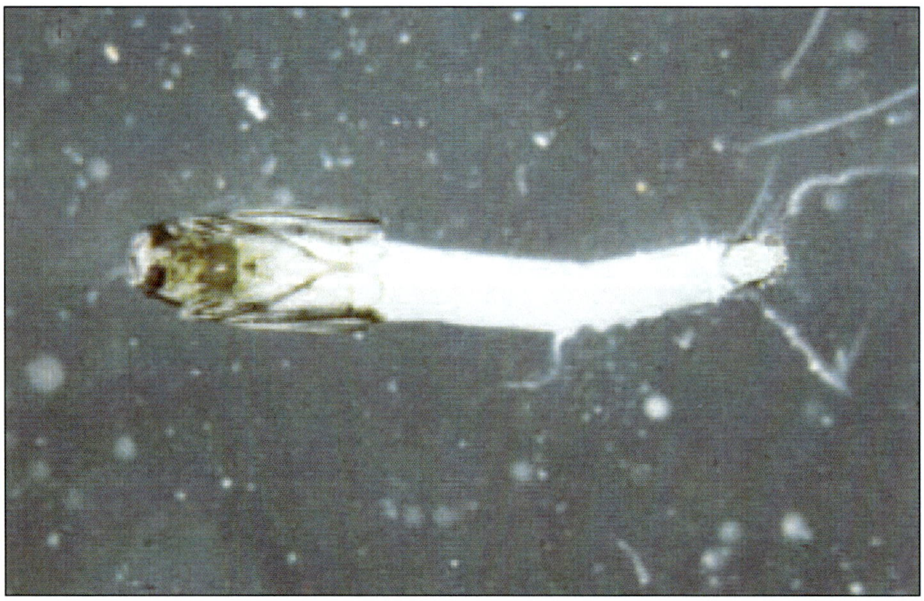

Source: Drs. J.H.M. van Lieverloo, Kiwa N.V. Research and Consultancy, Nieuwegein, The Netherlands.

Figure F-14 Chironomid. Pupa (Insecta, Chironomidae), ventral view, 1:20

Source: Drs. J.H.M. van Lieverloo, Kiwa N.V. Research and Consultancy, Nieuwegein, The Netherlands.

Figure F-15 Chironomid. Pupa (Insecta, Chironomidae), ventral view, 1:70

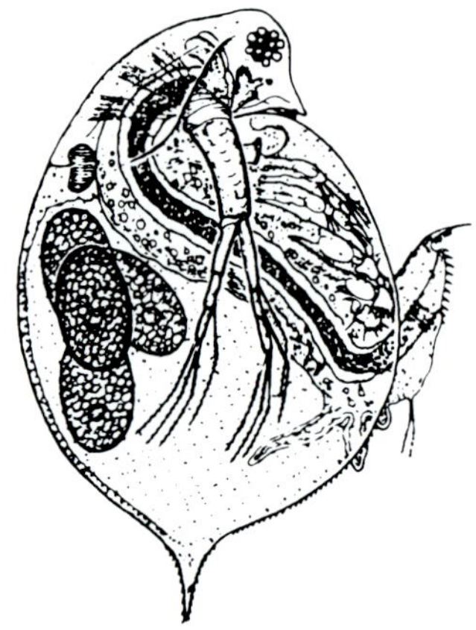

Source: H. Streble and D. Krauter. Poster Invertebrates in Drinking Water. Kosmos-Verlag, Stuttgart, Germany.

Figure F-16 *Daphnia* (2 mm)

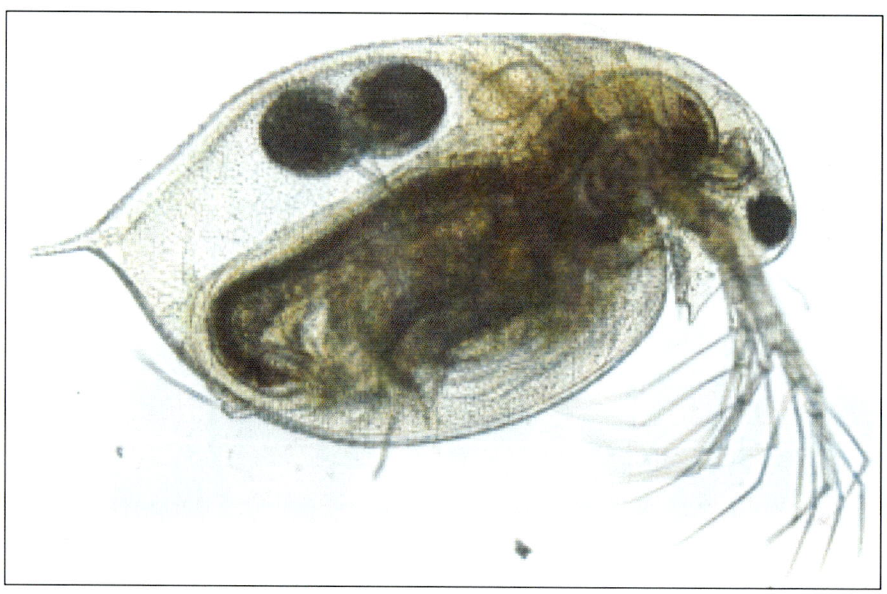

Source: Carolina Biological Supply Company.

Figure F-17 *Daphnia* sp. Water Flea (Crustacea)

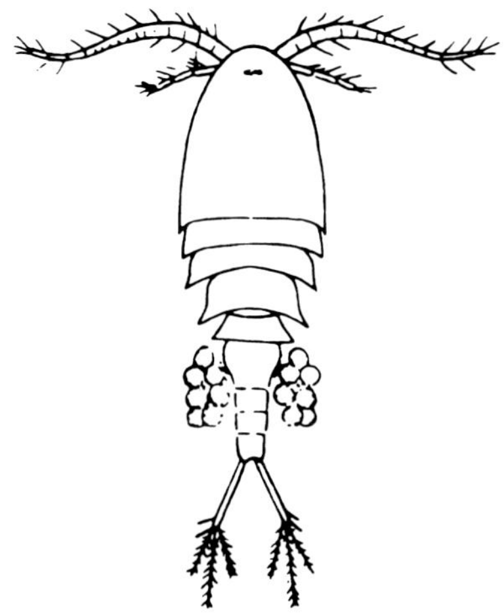

Source: *H. Streble and D. Krauter. Poster Invertebrates in Drinking Water. Kosmos-Verlag, Stuttgart, Germany.*

Figure F-18 *Cyclops* (2 mm)

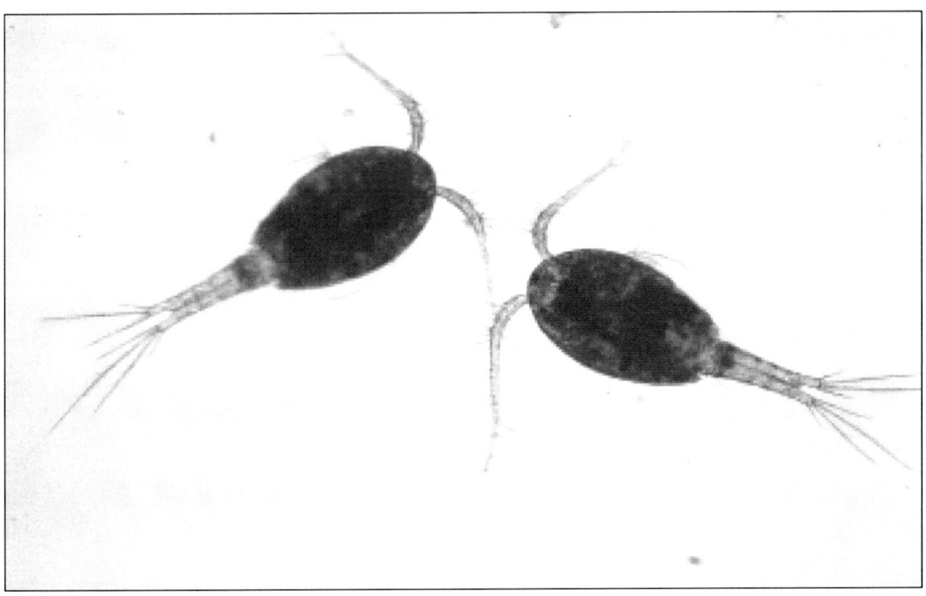

Source: *Carolina Biological Supply Company.*

Figure F-19 *Cyclops* sp. Water Flea (Crustacea)

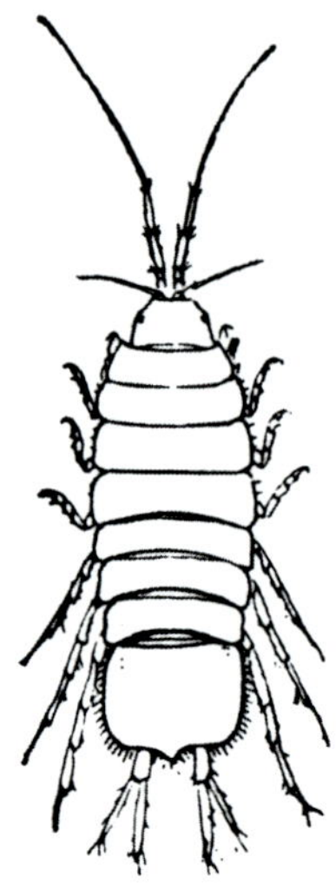

Source: Standard Methods for the Examination of Water and Wastewater.

Figure F-20 *Asellus aquaticus* (5 mm)

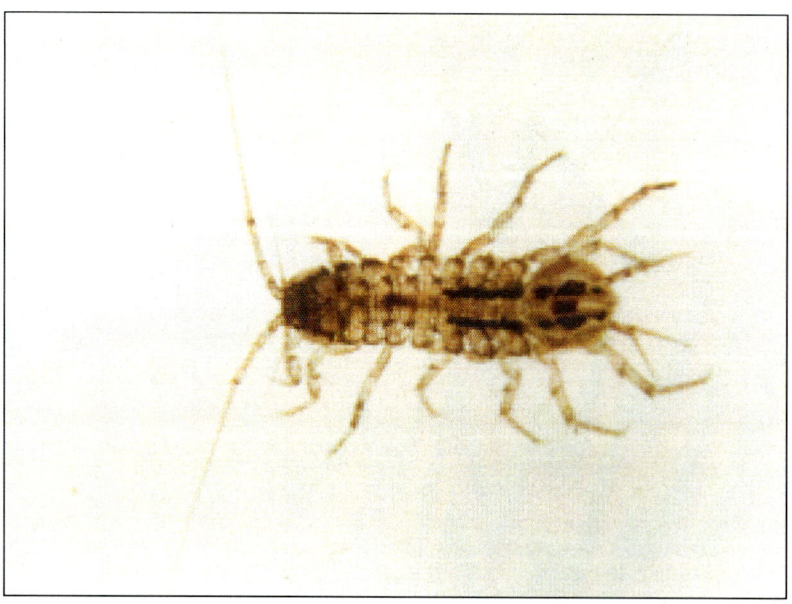

Source: Drs. J.H.M. van Lieverloo, Kiwa N.V. Research and Consultancy, Nieuwegein, The Netherlands.

Figure F-21 *Asellus aquaticus* (Crustacea, Isopoda), dorsal view, 1:5

Source: Drs. J.H.M. van Lieverloo, Kiwa N.V. Research and Consultancy, Nieuwegein, The Netherlands.

Figure F-22 *Asellus aquaticus* (Crustacea, Isopoda), lateral view, 1:20

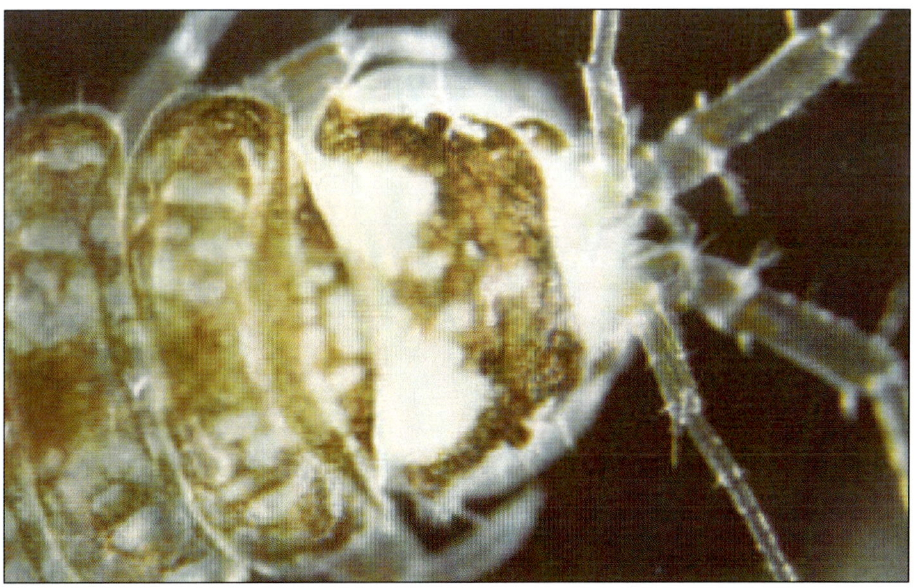

Source: Drs. J.H.M. van Lieverloo, Kiwa N.V. Research and Consultancy, Nieuwegein, The Netherlands.

Figure F-23 *Asellus aquaticus* (Crustacea, Isopoda), head, dorsal view, 1:25

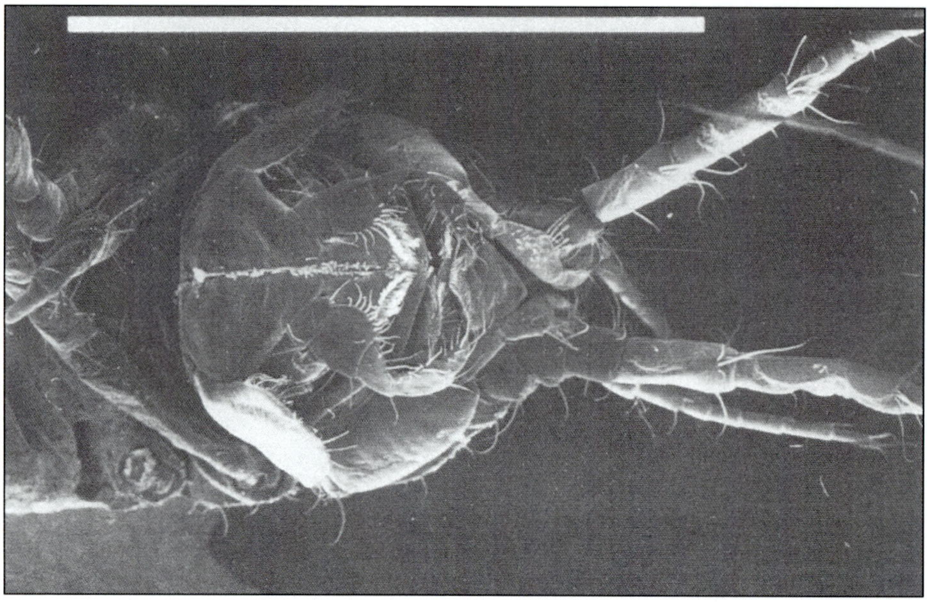

Source: *Drs. J.H.M. van Lieverloo, Kiwa N.V. Research and Consultancy, Nieuwegein, The Netherlands.*

Figure F-24 *Asellus aquaticus* (Crustacea, Isopoda), head, ventral view, 1:100

Source: *Drs. J.H.M. van Lieverloo, Kiwa N.V. Research and Consultancy, Nieuwegein, The Netherlands.*

Figure F-25 Faecal pellets of *Asellus aquaticus,* 1:4

Source: Drs. J.H.M. van Lieverloo, Kiwa N.V. Research and Consultancy, Nieuwegein, The Netherlands.

Figure F-26 Faecal pellets of *Asellus aquaticus* containing bitumen, 1:40

Source: Drs. J.H.M. van Lieverloo, Kiwa N.V. Research and Consultancy, Nieuwegein, The Netherlands.

Figure F-27 Faecal pellets of *Asellus aquaticus* containing iron-rust, 1:40

Source: K. Hancock, CHDiagnostic & Consulting Service Inc.

Figure F-28 Rotifer. *Monostyla,* 60×

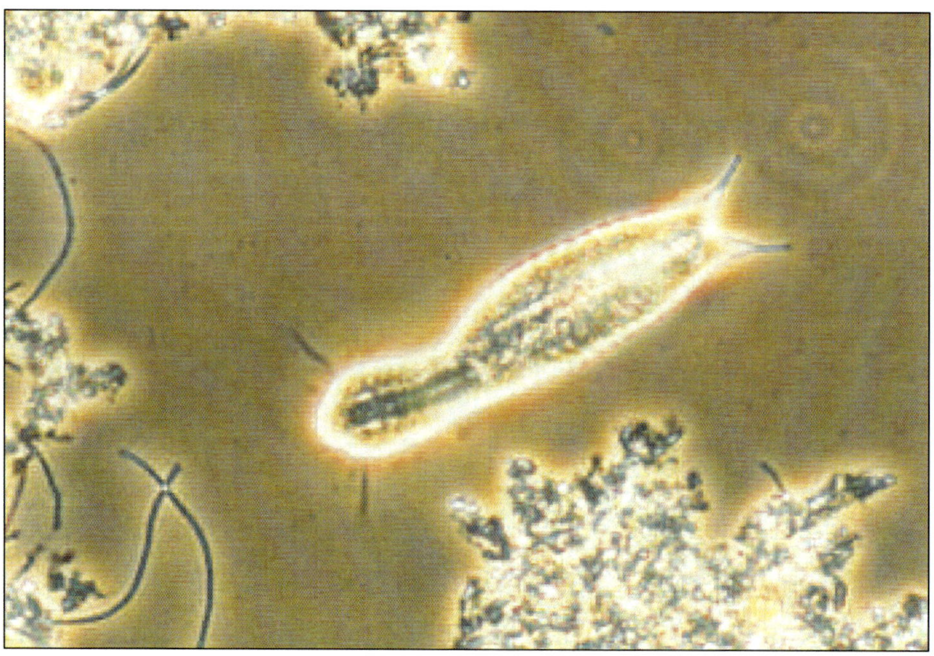

Source: M. Richard, PhD, RBD Inc., Ft. Collins, Colo.

Figure F-29 Rotifer. Gastrotrich (*Chaetonotus*), 200×

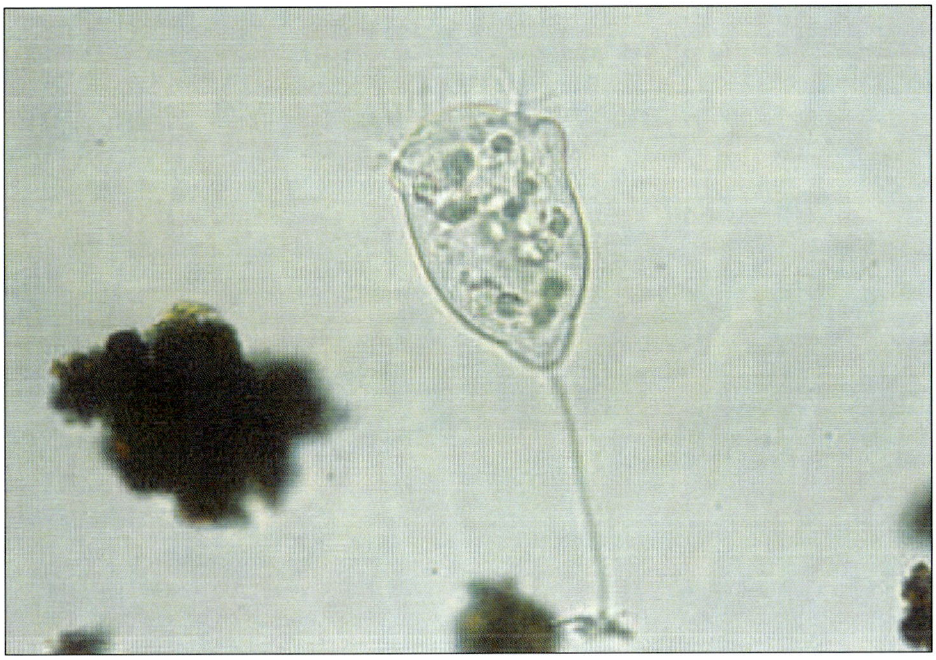

Source: K. Hancock, CHDiagnostic & Consulting Service Inc.

Figure F-30 Protozoan. Ciliate, *Vorticella,* unstained, bright-field microscopy, 120×

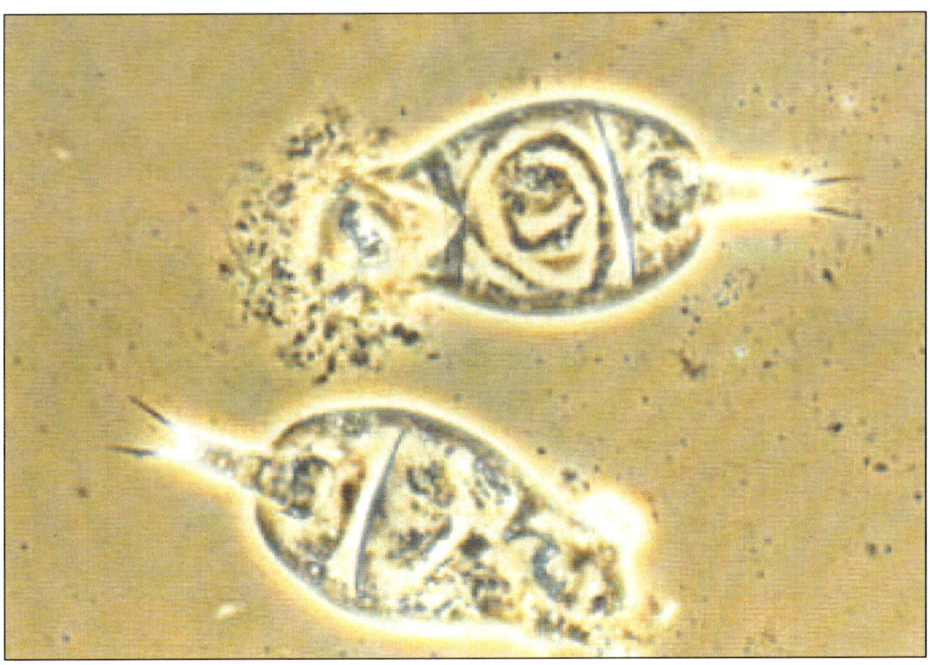

Source: M. Richard, PhD, RBD Inc., Ft. Collins, Colo.

Figure F-31 Rotifers. Rotifer, 200×

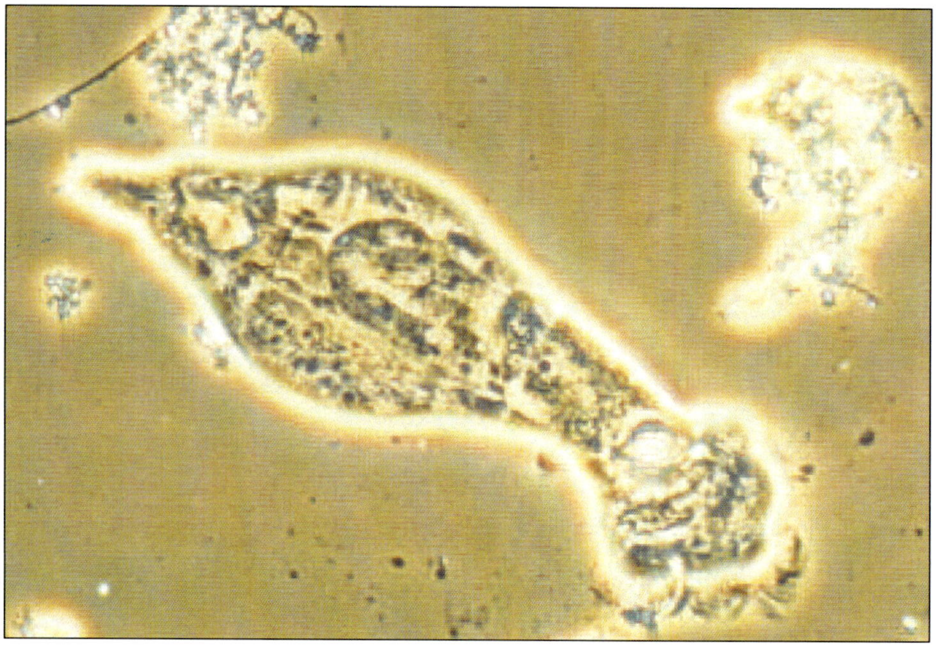

Source: M. Richard, PhD, RBD Inc., Ft. Collins, Colo.

Figure F-32 Gastrotrich, 200×

Source: C. Ramchran, University of Wisconsin Sea Grant Institute.

Figure F-33 Zebra mussels. Variations

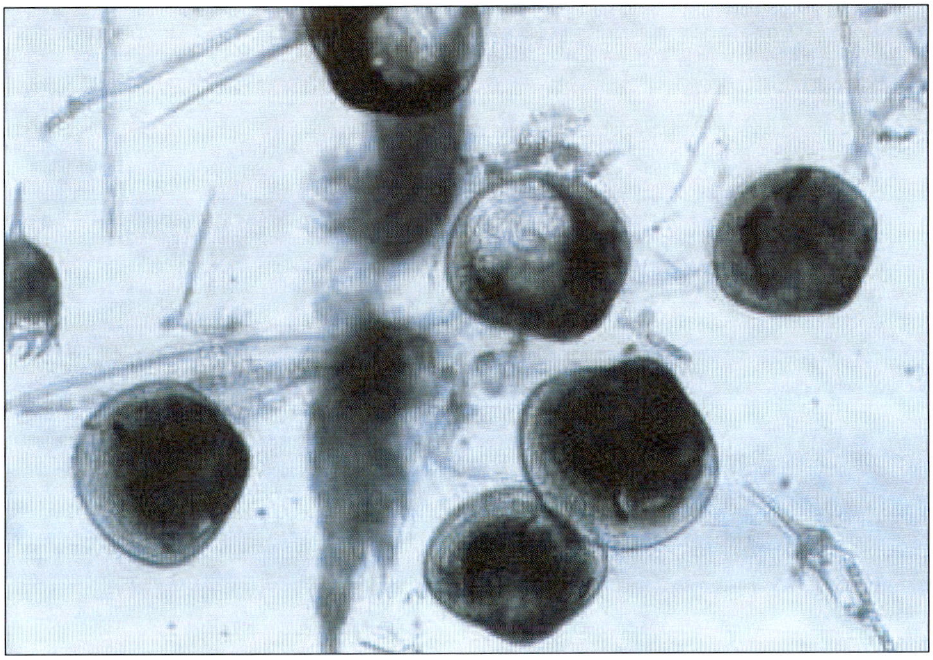

Source: University of Wisconsin Sea Grant Institute.

Figure F-34 Zebra mussel larvae showing some disintegration

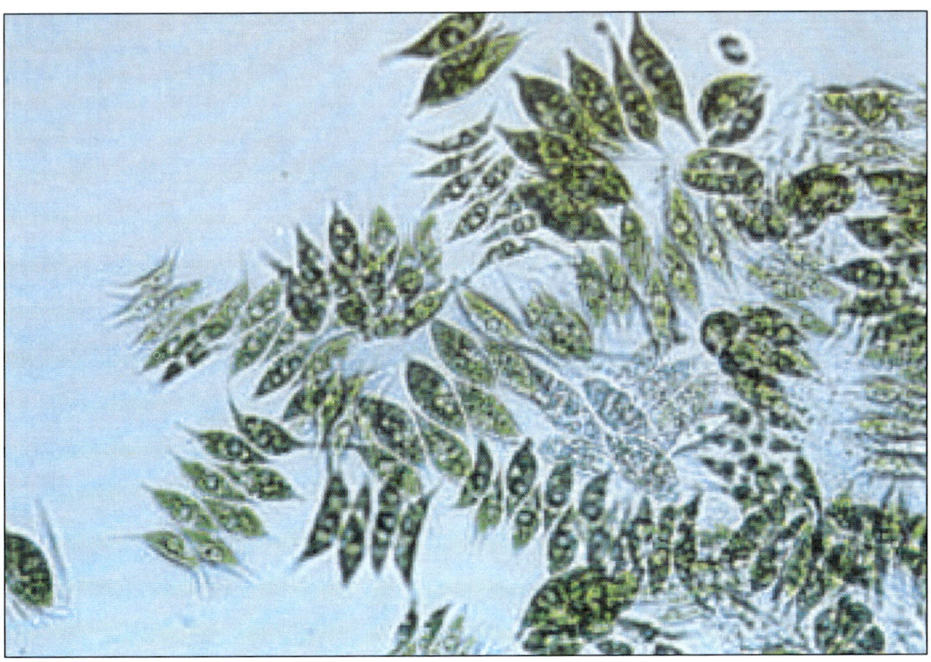

Source: K. Hancock, CHDiagnostic & Consulting Service Inc.

Figure F-35 Algae. Chlorophyte, *Scenedesmus,* unstained, bright-field microscopy, 120×

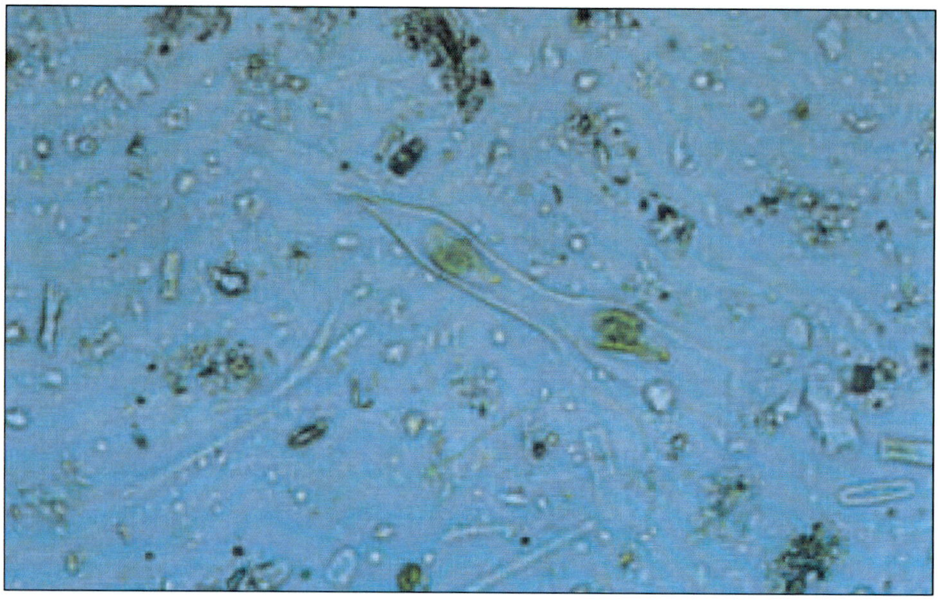

Source: K. Hancock, CHDiagnostic & Consulting Service Inc.

Figure F-36 Algae. Chrysophyte, *Dinobryon* (2 cells) and others, unstained, bright-field microscopy, 240×

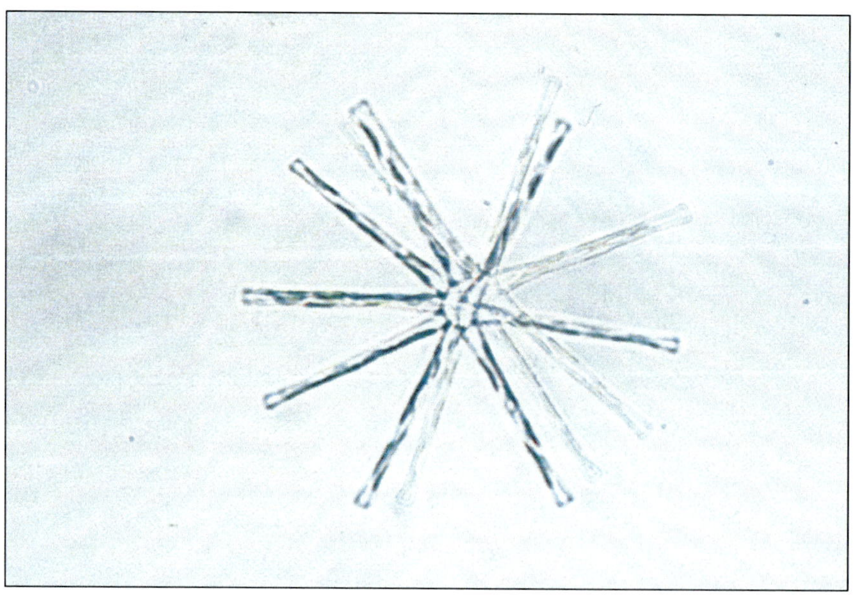

Source: K. Hancock, CHDiagnostic & Consulting Service Inc.

Figure F-37 Algae. Diatom, *Asterionella,* unstained, bright-field microscopy, 120×

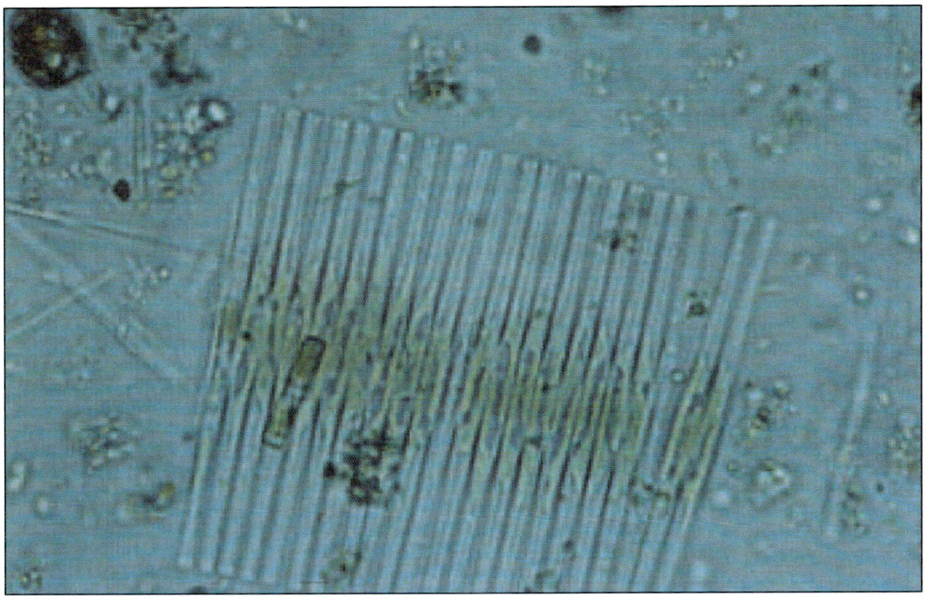

Source: K. Hancock, CHDiagnostic & Consulting Service Inc.

Figure F-38 Algae. Diatom, *Fragilaria,* unstained, bright-field microscopy, 240×

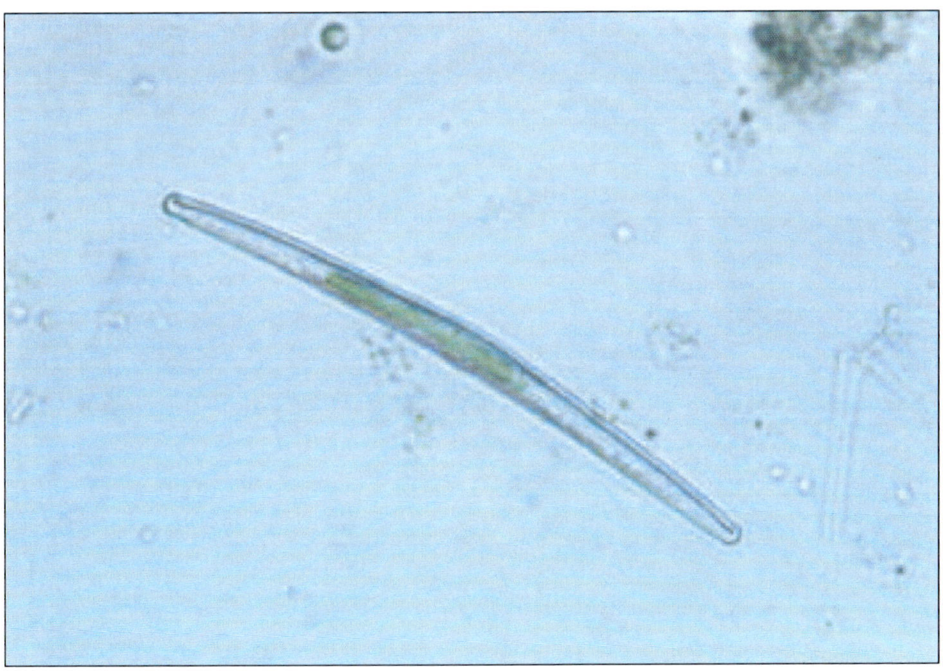

Source: K. Hancock, CHDiagnostic & Consulting Service Inc.

Figure F-39 Algae. Diatom, *Hannaea,* unstained, bright-field microscopy, 240×

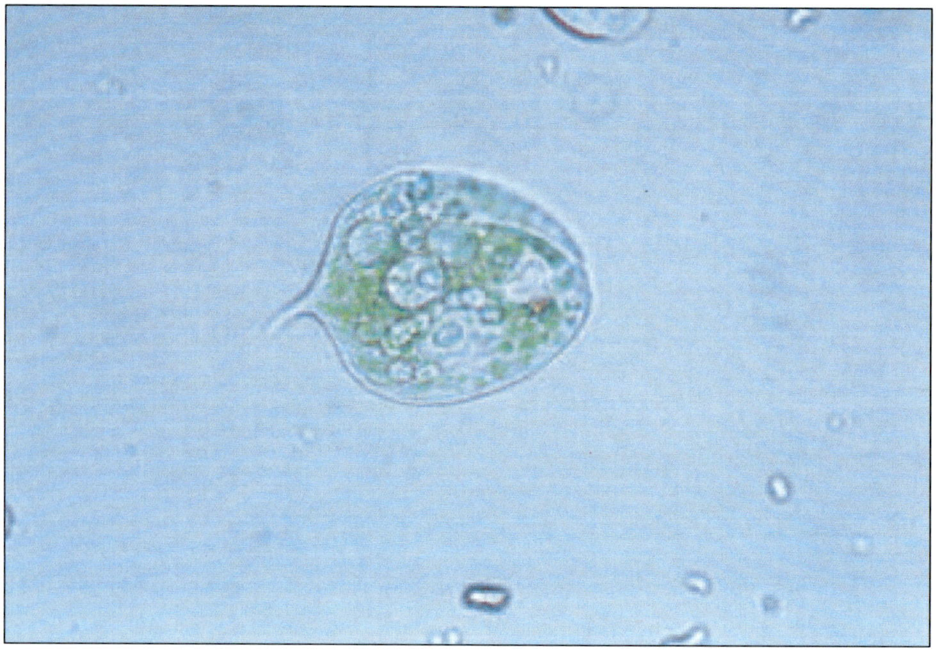

Source: K. Hancock, CHDiagnostic & Consulting Service Inc.

Figure F-40 Algae. Euglenophyte, *Phacus,* unstained, bright-field microscopy, 120×

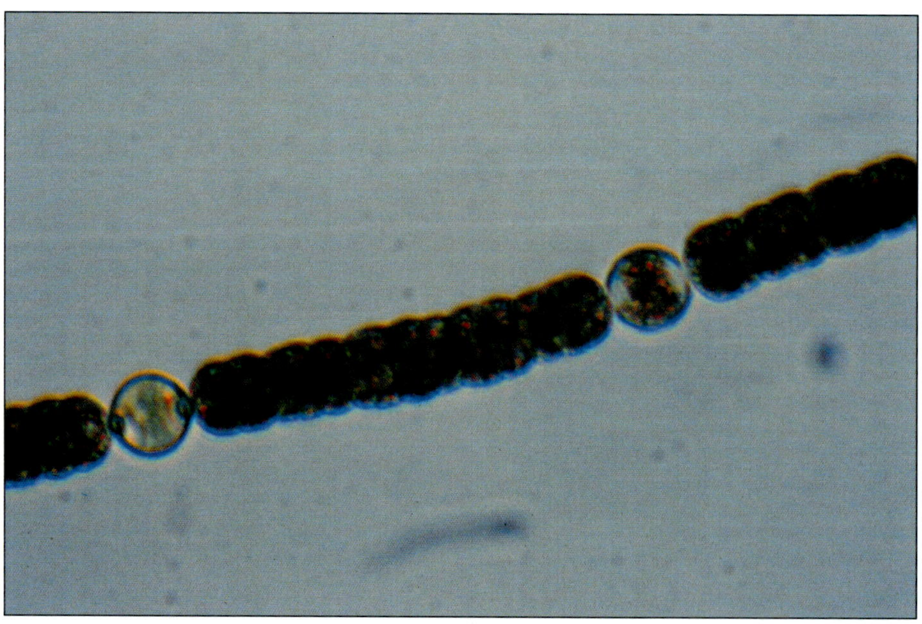

Source: G. Izaguirre, Metropolitan Water District of Southern California.

Figure F-41 Algae. *Anabaena Scherenietieri*, bright-field microscopy, 400×

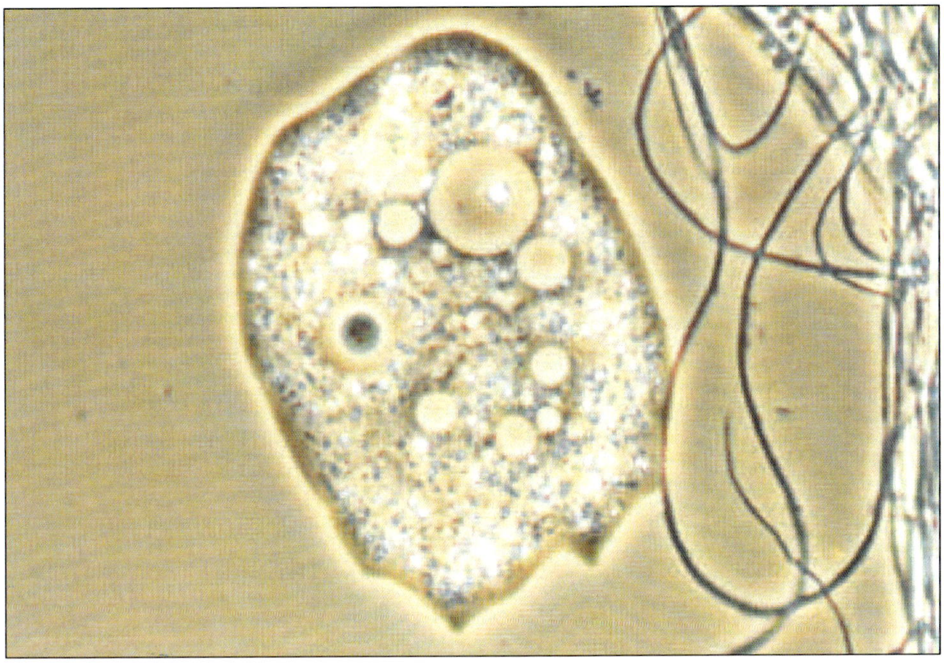

Source: M. Richard, PhD, RBD Inc., Ft. Collins, Colo.

Figure F-42 Protozoa. Amoeba, 200×

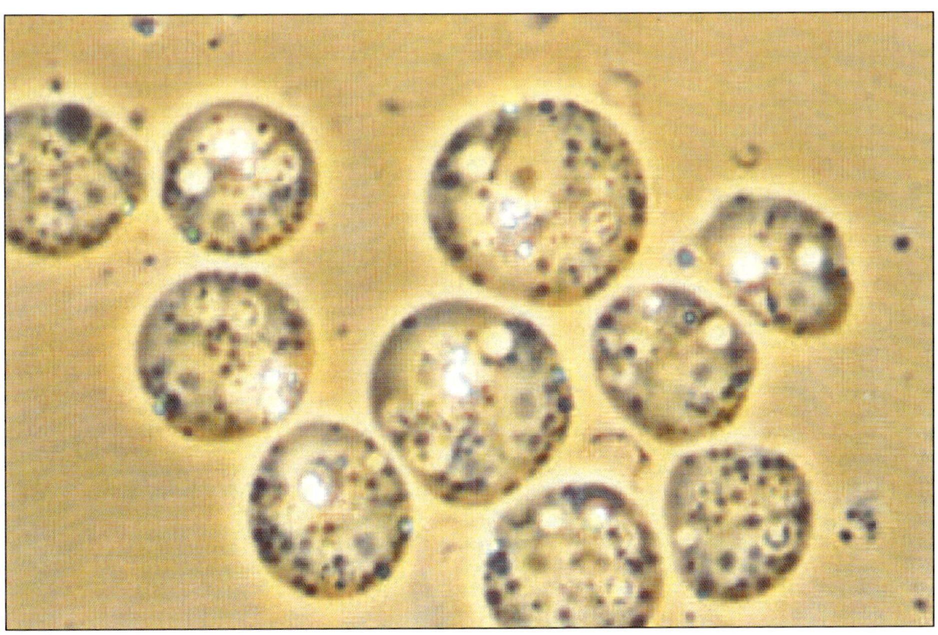

Source: M. Richard, PhD, RBD Inc., Ft. Collins, Colo.

Figure F-43 Protozoa. Flagellates, 1,000×

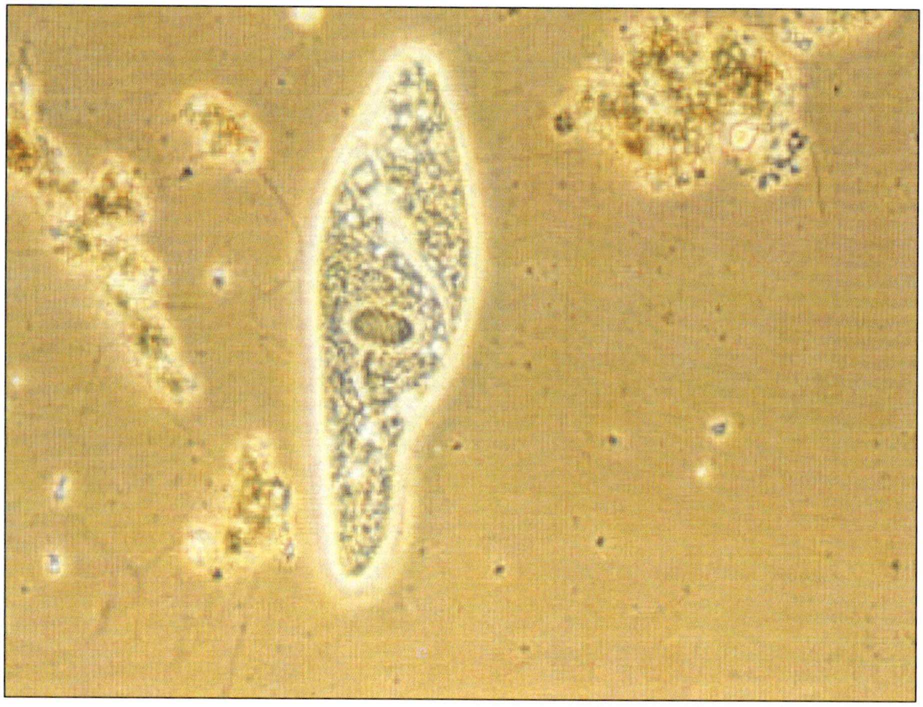

Source: M. Richard, PhD, RBD Inc., Ft. Collins, Colo.

Figure F-44 Protozoa. Free ciliate, 200×

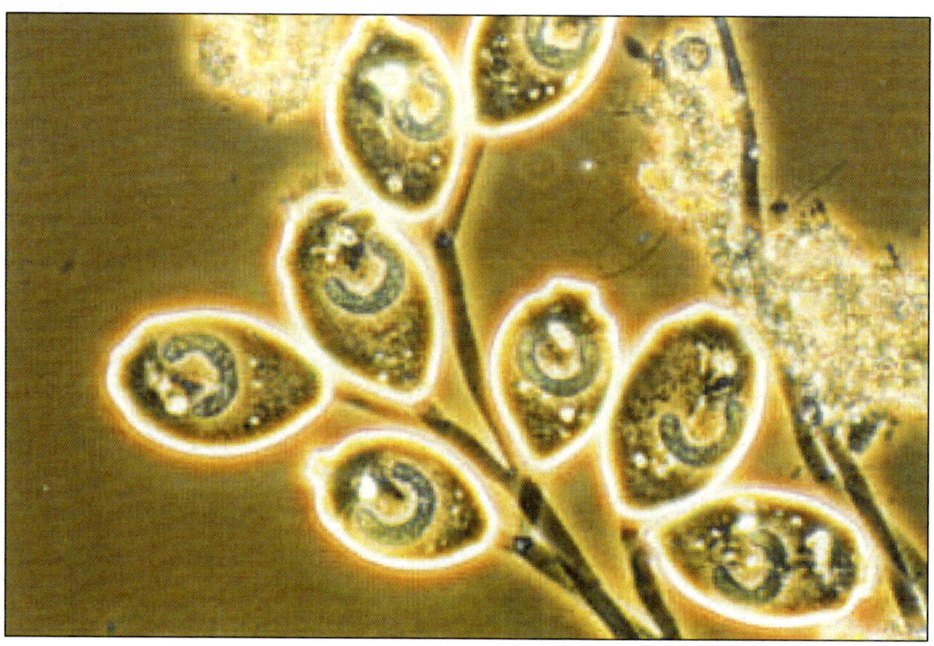

Source: M. Richard, PhD, RBD Inc., Ft. Collins, Colo.

Figure F-45 Protozoa. Stalked ciliate (stalks), 200×

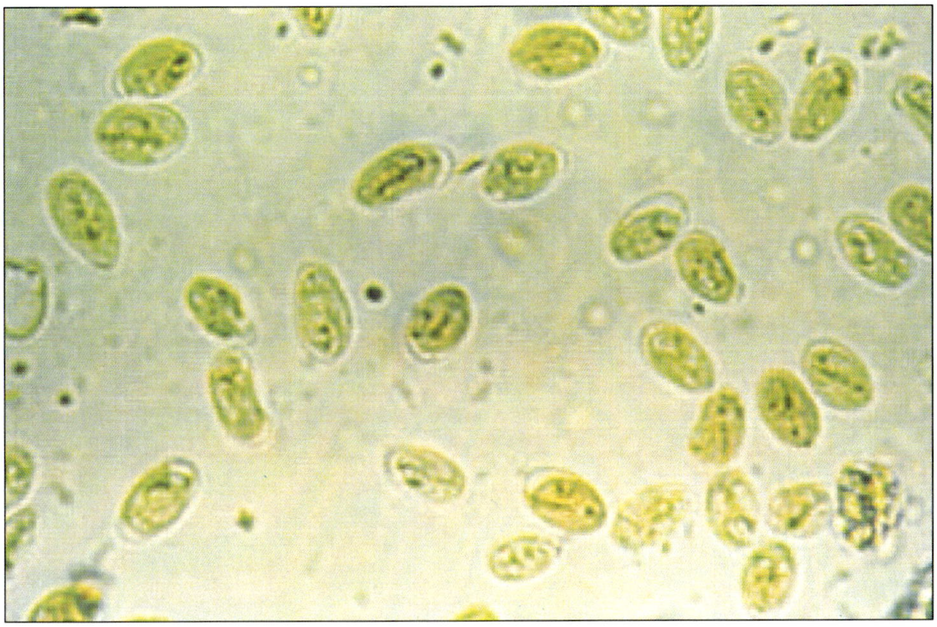

Source: CHDiagnostic & Consulting Service Inc. photo archive.

Figure F-46 Protozoa. *Giardia lamblia,* 240×

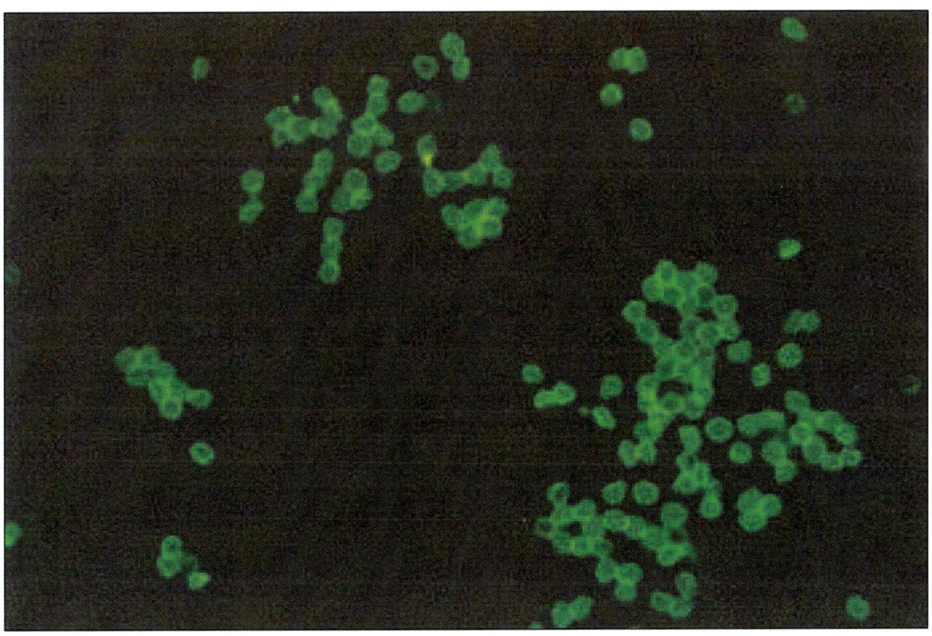

Figure F-47 Protozoa. *Cryptosporidium*

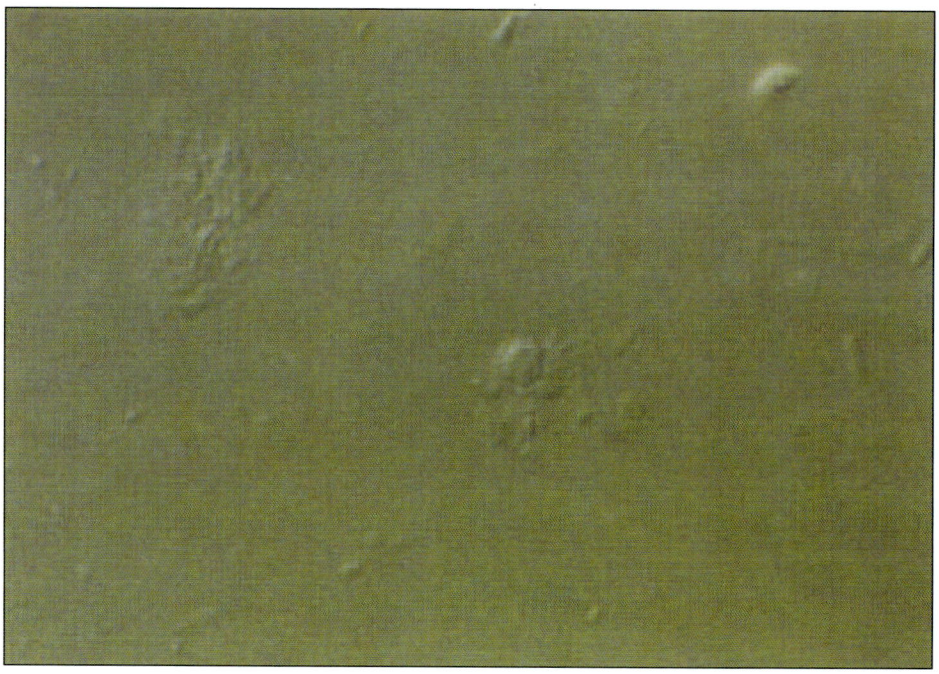

Source: *Scott Tighe, Analytical Services Inc.*

Figure F-48 *Cryptosporidium* oocyst exhibiting sporozoites, DIC microscopy, 1,000×

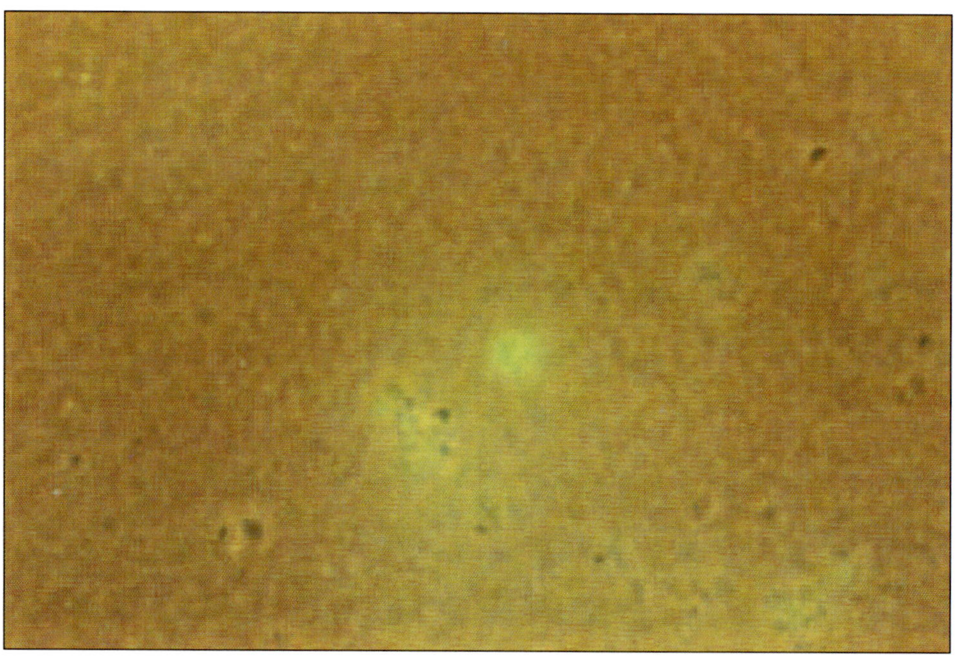

Source: *Scott Tighe, Analytical Services Inc.*

Figure F-49 *Cryptosporidium* oocyst exhibiting internal structures known as sporozoites, phase contrast, 1,000×

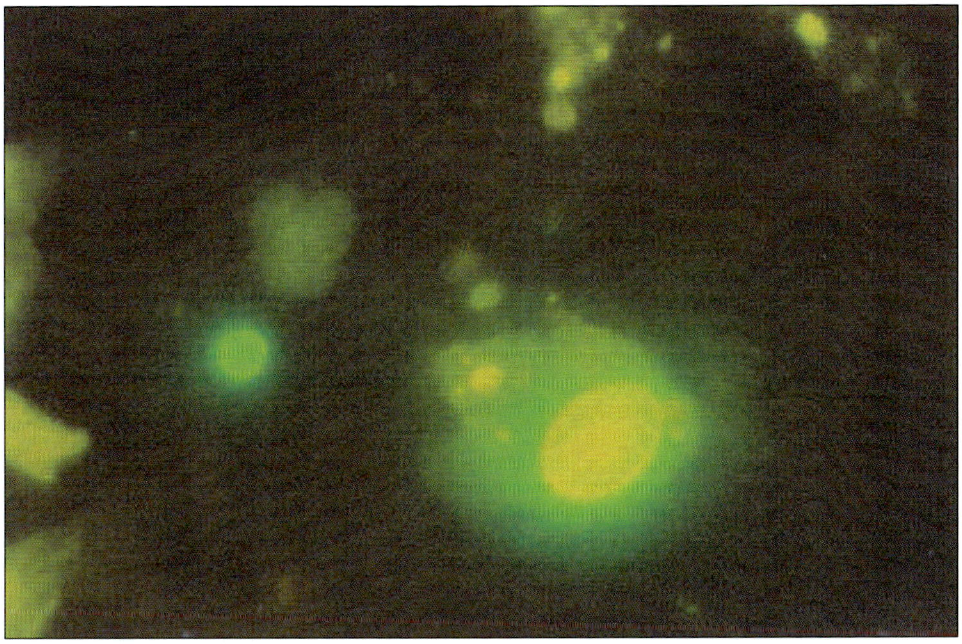

Source: Scott Tighe, Analytical Services Inc.

Figure F-50 *Giardia* cyst and *Cryptosporidium* oocyst, stained with fluorescent antibody stain, 1,000×

Source: Scott Tighe, Analytical Services Inc.

Figure F-51 *Giardia* cyst exhibiting internal structures—nuclei, axonemes, median bodies, DIC microscopy, 1,000×

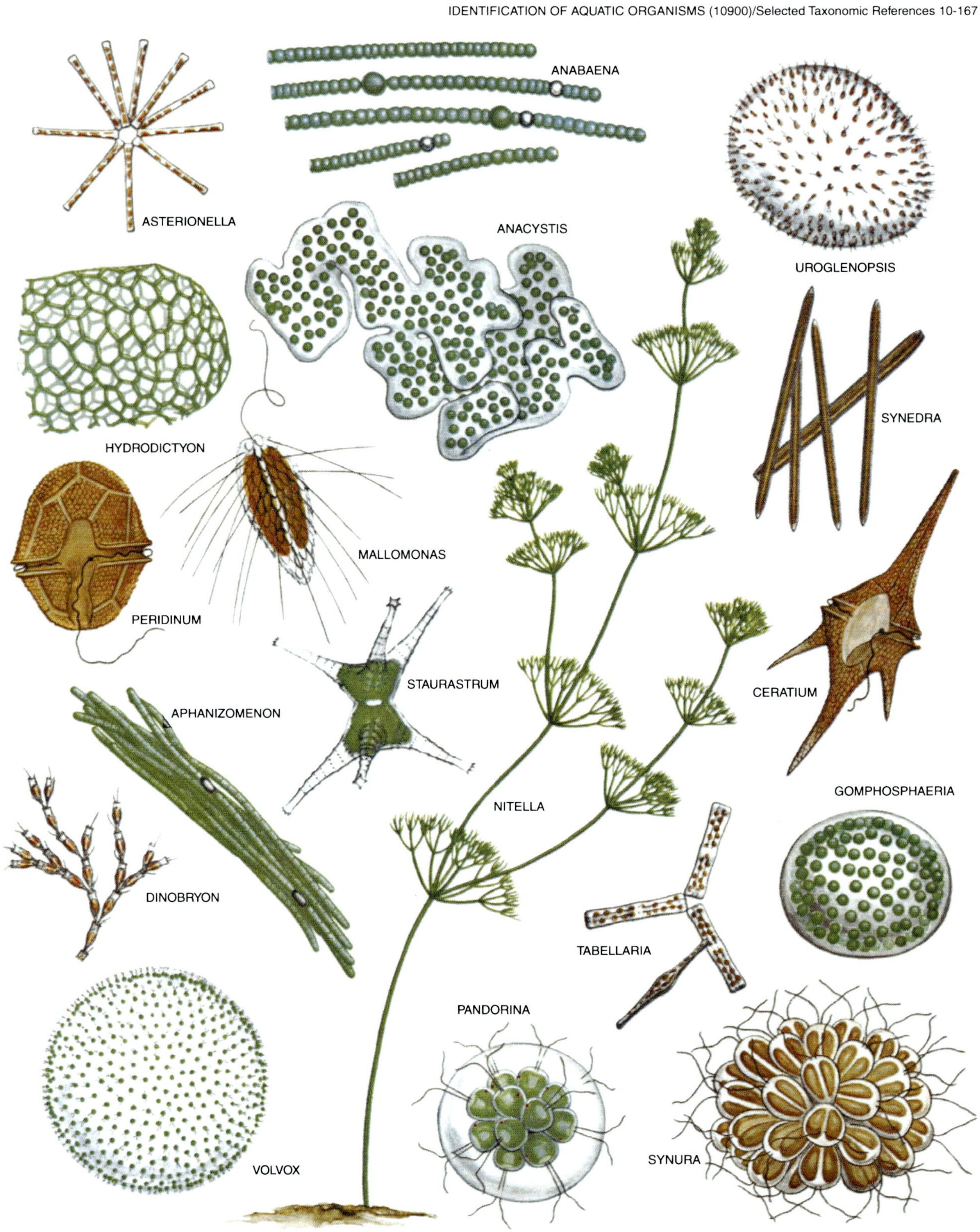

Source: Standard Methods for the Examination of Water and Wastewater.

Figure F-52 Taste-and-odor algae

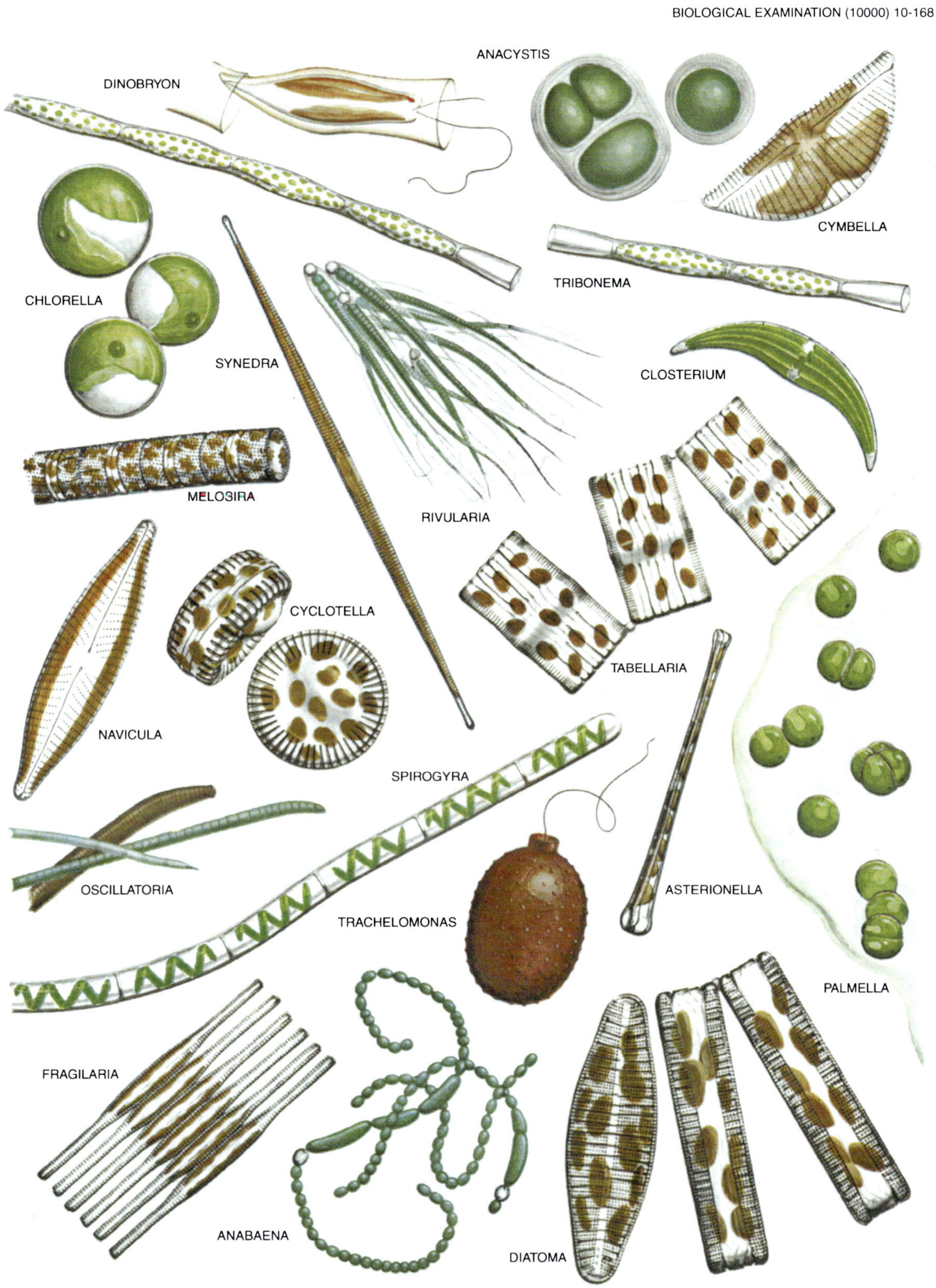

Source: Standard Methods for the Examination of Water and Wastewater.

Figure F-53 Filter-clogging algae

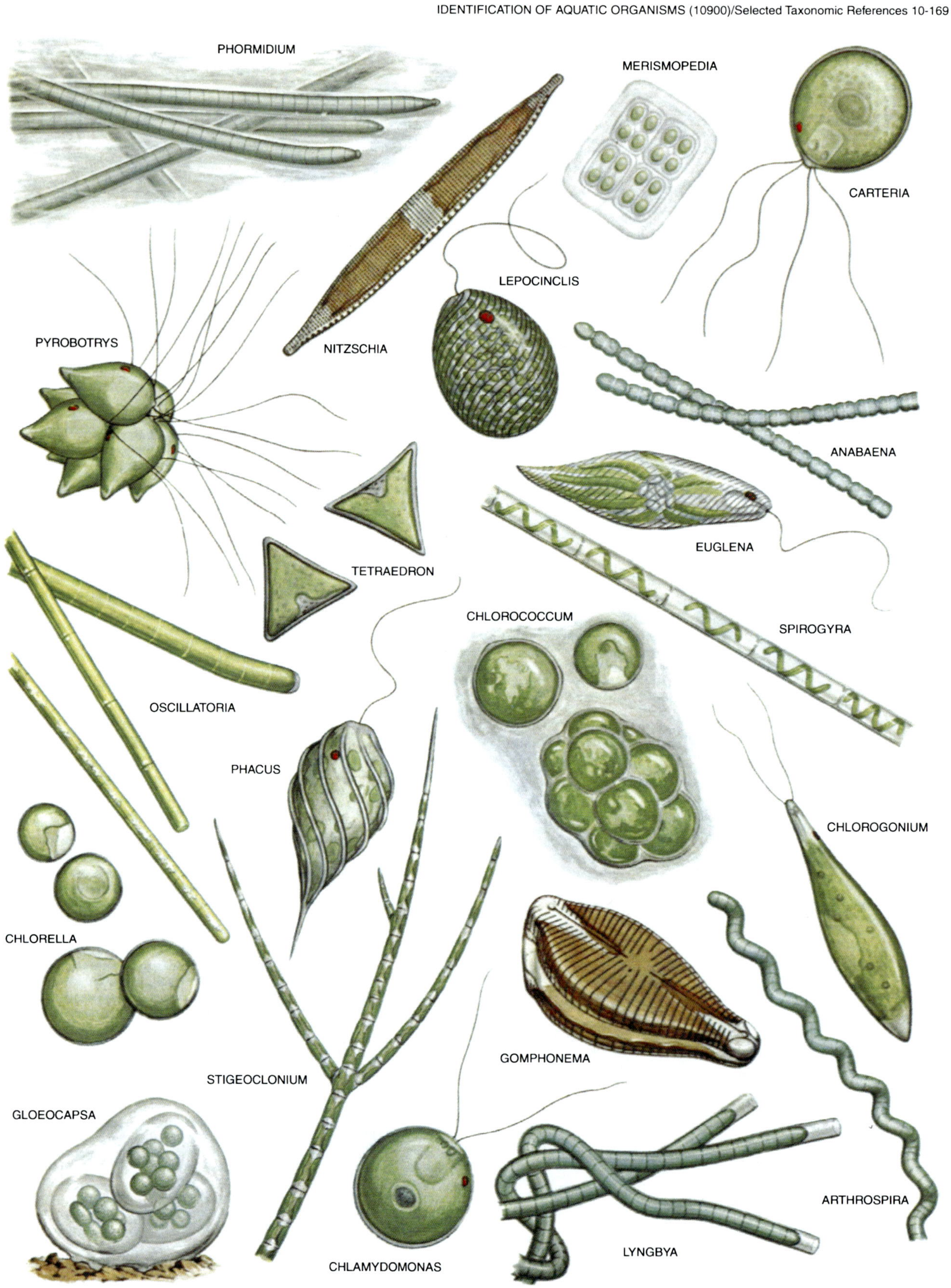

Source: Standard Methods for the Examination of Water and Wastewater.

Figure F-54 Polluted-water algae

Source: Standard Methods for the Examination of Water and Wastewater.

Figure F-55 Clean-water algae

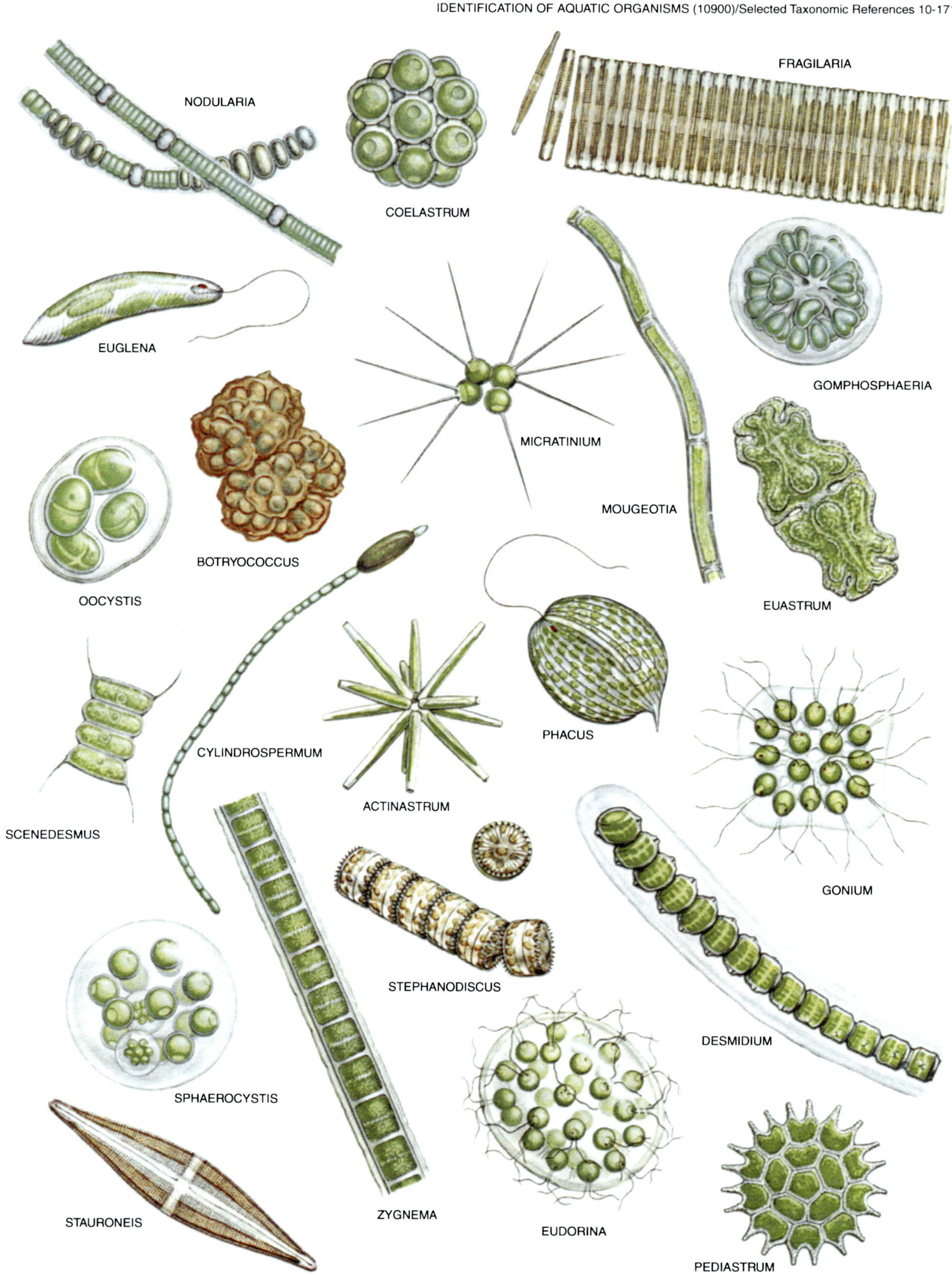

Source: Standard Methods for the Examination of Water and Wastewater.

Figure F-56 Plankton and surface algae

Source: Standard Methods for the Examination of Water and Wastewater.

Figure F-57 Reservoir algae

Source: Standard Methods for the Examination of Water and Wastewater.

Figure F-58 Wastewater-treatment-pond algae

Source: Standard Methods for the Examination of Water and Wastewater.

Figure F-59 Estuarine pollution algae

This page intentionally blank.

Abbreviations

AOB	ammonia-oxidizing bacteria
BOD	biological oxygen demand
CT	contact time
DIC	differential interference contrast
DO	dissolved oxygen
Eh	redox potential
IFA	immunofluorescent antibody
MCL	maximum contaminant level
MIB	2-methylisoborneol
MPA	microscopic particulate analysis
NOB	nitrite-oxidizing bacteria
PSP	paralytic shellfish poisoning
SRB	sulfate-reducing bacteria
TDH	total dynamic head
THM	trihalomethane
USEPA	US Environmental Protection Agency

This page intentionally blank.

Index

f. indicates a figure; *t.* indicates a table.

Abbreviations, 137
Acanthamoeba, 76
Actinomycetes, xv, 1
  and activated carbon adsorption, 5
  aerobic, 1
  anaerobic, 1
  and chlorination, 5
  control strategies, 5
  culture agars, 95–97
  egg albumin agar, 96
  favored conditions, 2
  filamentous form, 1, 2*f.*
  and flavor profile analysis, 4
  and flushing, 5
  and gas chromatography, 4
  and gas chromatography–mass spectroscopy, 4
  identification, 2–4, 3*f.*
  identification by cellular morphology, 3
  identification by culture characteristics, 3–4
  identification by direct examination of soils, water, and sludges, 2–3
  identification by indirect methods, 4
  isolation agar, 95
  M1B2 agar, 97
  microaerophilic, 1
  and ozone, 5
  and PEROXONE, 5
  significance, 4
  and sludge depth, 5
  sodium caseinate agar, 96
  spore form, 1–2
  starch agar, 97
  taste-and-odor compounds, 1, 2, 4
  and threshold odor test, 4
  troubleshooting, 88*t.*
  vegetative stage, 2, 3*f.*
Algae, xv, 57
  Anabaena Scherenietieri, 122*f.*
  Asterionella, 120*f.*
  attached (sessile), 57
  Bacillariophyceae (diatoms), 58, 62–63, 120*f.*, 121*f.*
  benthic, 57
  blocking filters, 58, 63
  blooms, 58, 60–61
  blue-green (Cyanophyta), xv, 58, 59, 60–61
  Chrysophyceae, 61–62, 63
  clean-water varieties, 131*f.*
  control strategies, 66
  copper sulfate as algicide, 66
  Cryptophyta, 64–65
  defined, 57
  Dinobryon, 120*f.*
  dinoflagellates (Pyrrhophyta), 64
  and elevated pH, 58
  estuarine-pollution varieties, 134*f.*
  Euglenophyta, 64, 122*f.*
  filter-clogging varieties, 129*f.*
  Fragilaria, 121*f.*
  green (Chlorophyta), 59–60
  Hannaea, 121
  as indicators of trophic state of water body, 58, 64
  obstruction of conveyance systems, 58
  and oxygen depletion, 58, 60–61
  oxygenation of water, 58 59
  periphyton, 57, 66
  Phacus, 122*f.*
  phyla, 59–65
  planktonic, 57, 66, 132*f.*
  polluted-water varieties, 130*f.*
  red (Rhodophyta), 65
  removal of minerals, 59
  reservoir varieties, 133*f.*
  role in food chain, 58
  Scenedesmus, 119*f.*
  significance, 57–59
  surface varieties, 132*f.*
  taste-and-odor-causing varieties, 128*f.*
  and tastes and odors, 57–58, 59–60, 64
  toxicity, 58, 61
  and trihalomethanes, 58
  troubleshooting, 88*t.*
  wastewater-treatment-pond varieties, 134*f.*
  Xanthophyceae, 62, 63
  yellow-green or golden-brown (Chrysop hyta), 61–64, 120*f.*
American Society of Testing and Materials, 99
American Water Works Association, Organisms in Water Committee, xv, 79
Amoebae, 69, 74, 123*f.*
Amphipoda, 43*f.*, 45, 46*f.*
Anabaena Scherenietieri, 122*f.*
ANSI/AWWA Standard C654 (Disinfection of Wells), 14
Arcella, 69, 60*f.*
Asellus, 43*f.*, 44, 45*f.*
Asellus aquaticus, 45*f.*, 112*f.*, 113*f.*
  faecal pellets, 114*f.*, 115*f.*
Asiatic clams, 51, 52
Asterionella, 120*f.*

Bacillariophyceae (diatoms), 58, 62–63, 120*f.*, 121*f.*
- *Asterionella*, 120*f.*
- *Fragilaria*, 121*f.*
- *Hannaea*, 121*f.*

Bacteria
- chemolithotrophic, 23
- heterotrophic, 23
- sheathed, 7
- *See also* Iron bacteria, Nitrifying bacteria, Sulfur bacteria

Balantidium coli, 75–76
Beggiatoa, 19, 20*f.*, 104*f.*
Bloodworms. *See* Midges
Blue-green algae, xv, 60–61
- blooms, 60–61
- contribution to fertility of rice fields, 59
- as indicator of eutrophic condition, 58
- and oxygen depletion, 58, 60–61
- as problem algae, 58, 60–61
- taste-and-odor problems, 60
- toxicity, 58, 61

Bosmina, 42, 42*f.*, 44*f.*, 46
Brown algae, 59

Caulobacter, 7, 8, 102*f.*
Chironomid larvae. *See* Midges
Chloramines, 23, 25, 26*f.*
Chlorine
- and actinomycetes, 5
- and algae, 66
- disinfection in control of nitrifying bacteria, 26, 27
- increasing chlorine-to-nitrogen weight ratio in control of nitrifying bacteria, 26
- and iron bacteria, 14
- superchlorination and nitrifying bacteria, 26
- and zebra mussels, 55

Chlorobium, 19
Chlorophyta, 59–60
Chromatium, 19
Chrysop hyta, 61–64, 120*f.*
Chrysophyceae, 61–62, 63
Ciliates, 69, 75, 124*f.*
- free, 124*f.*
- stalked, 124*f.*

Cladocera, 42–44, 42*f.*
Clonothrix, 7, 8
Coagulation
- optimization, 92*t.*
- and rotifers, 49

Coliforms, 8
Copepoda, 43*f.*, 44
Copper sulfate, 66
Corbicula fluminea. *See* Asiatic clams
Crustacea, 41, 42*f.*, 43*f.*
- *Amphipoda*, 43*f.*, 45, 46*f.*
- *Asellus*, 43*f.*, 44, 45*f.*
- *Asellus aquaticus*, 45*f.*, 112*f.*, 113*f.*, 114*f.*, 115*f.*
- *Bosmina*, 42, 42*f.*, 44*f.*, 46
- *Cladocera*, 42–44, 42*f.*
- control strategies, 45–46
- *Copepoda*, 43*f.*, 44
- *Cyclops*, 44, 46, 111*f.*
- *Daphnia*, 42, 42*f.*, 46, 110*f.*
- and disinfectants, 46
- and flushing, 46
- *Hyallela azteca*, 46*f.*
- Isopoda, 43*f.*, 44–45
- significance, 45
- taxonomic listing, 41
- troubleshooting, 88*t.*

Cryptodifflugia, 69
Cryptophyta, 64–65
Cryptosporidium, 71, 72*f.*, 74–75, 76, 77, 125*f.*–127*f.*
Cyanophyta, 60–61
Cyclops, 44, 46, 111*f.*
Cyclospora, 74, 77
Cyprus, 37
Cysts (oocysts, spores), 70

Daphnia, 42, 42*f.*, 46, 110*f.*
Delafield's hematoxylin, 74
Desulfobivrio, 19
Desulfobivrio desulfuricans, 20*f.*
Desulfotomaculum, 19
Diatoms, 58, 62–63, 120*f.*, 121*f.*
- *Asterionella*, 120*f.*
- *Fragilaria*, 121*f.*
- *Hannaea*, 121

Dinobryon, 120*f.*
Dinoflagellates, 64
Disinfection
- in control of nitrifying bacteria, 26, 27
- and crustacea, 46
- optimization, 93*t.*
- *See also* Chloramines, Chlorine, Ozone

Distribution system
- controlling iron bacteria in, 15
- optimization, 93*t.*

Dreissena polymorpha. *See* Zebra mussels

Encephalitozoon intestinalis, 76
England, 37, 38
Entamoeba coli, 74
Entamoeba histolytica, 74, 77
Enterocytozoon bieneusi, 76
Essex (England) Water Company, 38
Euglena, 71
Euglenophyta, 64, 122*f.*
- *Phacus*, 122*f.*

Evaluation and Restoration of Water Supply Wells, 14

Filtration
  algae blocking filters, 58, 63
  and iron bacteria, 15
  membrane filters in identification of nematodes, 31
  nematodes and rapid sand filtration, 33
  nitrifying bacteria and biologically active filtration, 23, 25
  optimization, 92*t*.–93*t*.
  and rotifers, 49
  and zebra mussels, 55
Flagellates, 69, 123*f*.
Flavor profile analysis, 4
Flocculation
  optimization, 92*t*.
  and rotifers, 49
Flushing
  and actinomycetes, 5
  and crustacea, 46
  and nematodes, 32
  and nitrifying bacteria, 26
Fragilaria, 121*f*.

Gallionella, 7, 8
Gallionella ferruginea, 8, 9*f*., 10*f*.
Gas chromatography, 4
Gas chromatography–mass spectroscopy, 4
Gaston County, North Carolina, 38
Gastrotrich, 116*f*., 118*f*.
Geosmin, 60
Giardia, 71–73, 72*f*., 76, 77, 125*f*., 127*f*.
Golden-brown algae, 61–64
Granular activated carbon, 5
Great Lakes, 51
Green algae, 59–60
Greensand, 14

Hannaea, 121
Hoffman modulation, 73
Hyallela azteca, 46*f*.
Hydrogen peroxide. *See* PEROXONE
Hyphomicrobium, 7, 8

Iron bacteria, xv, 7, 8*f*.
  with appendages, 7
  biofouling sequence, 11, 11*f*., 12*f*., 13*f*.
  Caulobacter, 7, 8, 102*f*.
  Clonothrix, 7, 8
  coliforms, 8
  control in distribution system, 15
  control in filters, 15
  control strategies, 12–15
  controlling fouling in wells, 14
  culturing, 8–10
  and environmental factors, 8
  filamentous or stalked types, 10, 10*f*.
  Gallionella, 7, 8
  Gallionella ferruginea, 8, 9*f*., 10*f*.
  heterotrophic iron-precipitators, 8
  Hyphomicrobium, 7, 8
  identification, 8–11
  identification by direct examination, 10–11
  interaction with iron, 7
  Leptothrix, 7, 8, 102*f*.
  monitoring, 13, 15
  and pipe corrosion and tuberculation, 15
  presence/absence test method, 99–100
  preventing fouling in wells, 12–14
  Pseudomonas, 8, 103*f*.
  quantity test method, 100
  sheathed bacteria, 7
  significance, 11
  Sphaerotilus, 7, 8
  Sphaerotilus natans, 8, 9*f*.
  Thiobacillus, 7, 8, 103*f*.
  troubleshooting, 88*t*.–89*t*.
  varieties, 7–8
Isopoda, 43*f*., 44–45
Isospora, 74, 75, 77
Isospora hominis, 75

Larvae, xv
Legionella, 76
Leptothrix, 7, 8, 102*f*.
Lowell (Indiana) Water Utility, 37, 38–39
Lugol's iodine, 73

MIB, 60
Michigan State University, Department of Entomology, 39
Microsporidians, 76
Midges, 35, 36*f*., 37*f*.
  and cationic polymers, 39
  control strategies, 38–39
  eggs, 36
  and fine-mesh screen, 39
  identification, 36–37
  as indicators of ecosystem health, 35
  as indicators of toxic substances, 35–36
  infestations, 37–38
  larvae, 36
  life cycle, 36
  lifespan, 35
  limiting food supply, 39
  Paratanytarsus grimmii, 37
  parthenogenetic reproduction, 38, 39*f*.
  pupae, 37, 37*f*., 108*f*., 109*f*.
  and pyrethrins (insecticide), 38
  significance, 37–38
  troubleshooting, 88*t*.
Missouri, 38

Nematodes, 29, 30*f*., 31*f*., 32*f*., 107*f*.
  control strategies, 32–33
  feeding, 29
  filters in identification, 31
  and flushing, 32
  habitats, 30–31

identification, 31
larva, 107*f.*
life cycle, 29–30
locomotion, 29, 30*f.*
and pathogens, 32
physical characteristics, 29
and rapid sand filtration, 33
and settling, 33
significance, 32
and source waters, 30–31
troubleshooting, 89*t.*
Nitrification, 23
Nitrifying bacteria, 23, 24*f.*
and ammonia, 23, 25
ammonia-oxidizing bacteria, 23–24
and biologically active filtration, 23, 25
chemolithotrophic, 23
and chloramines, 23, 25, 26*f.*
and chlorine disinfection, 26, 27
control strategies, 26–27
and decreasing detention time, 26–27
and flushing, 26
habitats, 24
identification, 25
and increasing chlorine-to-nitrogen weight ratio, 26
monitoring, 27
nitrite-oxidizing bacteria, 23–24
Nitrobacter, 24, 25, 106*f.*
Nitrococcus, 24
and nitrogen cycle, 24
Nitrosococcus, 24
Nitrosomonas, 24, 25, 106*f.*
Nitrosovibrio, 24
Nitrospina, 24
Nitrospira, 24
and pH, 24
shapes, 23
significance, 25
and superchlorination, 26
troubleshooting, 89*t.*
and ultraviolet irradiation, 24
Nitrobacter, 24, 25, 106*f.*
as nitrite-oxidizing bacteria, 24, 25
Nitrococcus, 24
Nitrosococcus, 24
as ammonia-oxidizing bacteria, 24
Nitrosomonas, 24, 25, 106*f.*
as ammonia-oxidizing bacteria, 24, 25
Nitrosovibrio, 24
as ammonia-oxidizing bacteria, 24
Nitrospina, 24
Nitrospira, 24
Normarski differential interference contrast microscopy, 73
Nuisance Organisms survey, xv, 79
form, 80
results, 81–85

Optimization of conventional water treatment, 91, 92*t.*–93*t.*
Ozone
and actinomycetes, 5
and zebra mussels, 55
See also PEROXONE

Paralytic shellfish poisoning, 58
Paramecium, 69–70, 70*f.*
Paratanytarsus grimmii, 37
PEROXONE, 5
Phacus, 122*f.*
Phaeophyta, 59
Phytoplankton, 57, 66
Plankton, 57, 66, 132*f.*
Powdered activated carbon, 5
Protozoa, 69
Acanthamoeba, 76
amoebae, 69, 74, 123*f.*
Arcella, 69, 60*f.*
Balantidium coli, 75–76
ciliates, 69, 75, 124*f.*
commensal, 70
control strategies, 77
Cryptodifflugia, 69
Cryptosporidium, 71, 72*f.*, 74–75, 76, 77, 125*f.*–127*f.*
Cyclospora, 74, 77
cysts (oocysts, spores), 70
Encephalitozoon intestinalis, 76
Entamoeba coli, 74
Entamoeba histolytica, 74, 77
enteric, 70
Enterocytozoon bieneusi, 76
Euglena, 71
flagellates, 69, 123*f.*
free ciliate, 124*f.*
free-living, 69–70, 71
Giardia, 71–73, 72*f.*, 76, 77, 125*f.*, 127*f.*
identification, 71–76
Isospora, 74, 75, 77
Isospora hominis, 75
and meningitis, 69
microsporidians, 76
Paramecium, 69–70, 70*f.*
parasitic, 69, 70
Quadrulella, 69
Sarcomastigophora, 74
significance, 76
staining, 71, 73, 74, 75–76
stalked ciliate, 124*f.*
taxonomic listing, 71, 73*t.*
Toxoplasma, 74, 75, 77
Toxoplasma gondii, 75
trophozoites, 70
troubleshooting, 89*t.*
Prymnesiophytes, 59
Pseudomonas, 8, 103*f.*

Pyrethrins, 38
Pyrrhophyta, 64

Quadrulella, 69

Rapid sand filtration, 33
Red algae, 65
Rhodophyta, 65
Rotifers, 47, 48*f.*, 49*f.*, 116*f.*, 117*f.*
  and coagulation, flocculation, and filtration, 49
  control strategies, 49–50
  and dissolved oxygen, 47
  Gastrotrich, 116*f.*, 118*f.*
  identification, 47–48, 48*f.*
  as indicators of surface water contamination of groundwater, 49
  mastax, 48, 49*f.*
  shapes and "armor," 48
  significance, 49
  in toxicity testing, 47
  troubleshooting, 89*t.*
  Vorticella, 117*f.*
  and water quality, 47
  wheel-like circles of cilia, 47, 48*f.*
Roundworms. *See* Nematodes

Salmonella, 32
Sandoz accident (Switzerland), 35
Sarcomastigophora, 74
Scenedesmus, 119*f.*
Sedimentation, 49*t.*
Shigella, 32
Sludge depth, 5
Source waters
  and nematodes, 30–31
  optimization, 92*t.*
Sphaerotilus, 7, 8
Sphaerotilus natans, 8, 9*f.*
SRB. *See* Sulfur bacteria
Standard Methods for the Examination of Water and Wastewater
  on actinomycetes, 4
  on iron bacteria, 8, 10, 99–100
  on midges, 36
  on nematodes, 31, 33
  on sulfur bacteria, 19–20
Standard Test Method for Iron Bacteria in Water and Water-Formed Deposits, 99
Streptomycetes, 1
Sulfate-reducting bacteria. *See* Sulfur bacteria
Sulfur bacteria, xv, 7, 17
  Beggiatoa, 19, 20*f.*, 104*f.*
  black slimes, 19–21
  Chlorobium, 19
  Chromatium, 19
  colorless, 19
  control strategies, 21–22
  and corrosion, 17–19, 20
  Desulfobivrio, 19
  Desulfobivrio desulfuricans, 20*f.*
  Desulfotomaculum, 19
  and dissolved oxygen, 21–22
  hydrogen sulfide gas, 17
  identification, 19–20
  and oxygen, 17, 19
  primary indicators, 19
  and pumping, 21, 21*f.*, 22*f.*
  purple and green, 19
  and "rotten egg" smell, 17
  significance, 20–21
  and slimes, 17, 18*f.*, 21
  sulfate-reducing, 19
  sulfur-oxidizing, 19, 20*f.*
  sulfur-reducing, 19
  Thiobacillus, 19, 104*f.*
  Thiopedia, 19, 105*f.*
  Thiothrix, 19, 105*f.*
  troubleshooting, 89*t.*
  and tubercles, 17, 18*f.*, 21

Tastes and odors
  and actinomycetes, 1
  and algae, 57–58
  and blue-green algae, 60
  and dinoflagellates (Pyrrhophyta), 64
  and green algae, 59–60
  and sulfur bacteria, 17
  and zebra mussels, 53
Thiobacillus, 7, 8
  as sulfur bacteria, 19, 104*f.*
Thiobacillus ferro oxidans, 103*f.*
Thiopedia, 19, 105*f.*
Thiothrix, 19, 105*f.*
Threshold odor test, 4
Toxoplasma, 74, 75, 77
Toxoplasma gondii, 75
Trophozoites, 70
Troubleshooting guide, 87, 88*t.*–89*t.*
2-methylisoborneol (MIB), 60

US Environmental Protection Agency, 31

Vorticella, 117*f.*

Washington County, North Carolina, 38

Xanthophyceae, 62, 63

Yellow-green algae, 61–64

Zebra mussels, 51, 53*f.*, 118*f.*
  and antifouling coatings, 55
  chemical control, 55
  and chlorine, 55
  control strategies, 54–56
  identification, 52

introduction to North America, 51
life cycle, 51, 52*f.*
masses, 52, 53*f.*
mechanical control, 56
and oxygen deprivation, 55
and ozone, 55
physical control, 55
and sand infiltration beds, 55
scraping, 56
screening, 56
significance, 53–54
and taste-and-odor problems, 53
thermal control, 55
troubleshooting, 89*t.*
veliger larvae, 51, 52, 54*f.*, 56, 119*f.*
and water intake structures, 53, 54*f.*
Zooplankton, 57

AWWA Manuals

M1, *Principles of Water Rates, Fees, and Charges,* Fifth Edition, 2000, #30001PA

M2, *Instrumentation and Control,* Third Edition, 2001, #30002PA

M3, *Safety Practices for Water Utilities,* Sixth Edition, 2002, #30003PA

M4, *Water Fluoridation Principles and Practices,* Fourth Edition, 1995, #30004PA

M5, *Water Utility Management,* First Edition, 1980, #30005PA

M6, *Water Meters—Selection, Installation, Testing, and Maintenance,* Fourth Edition, 1999, #30006PA

M7, *Problem Organisms in Water: Identification and Treatment,* Third Edition, 2003, #30007PA

M9, *Concrete Pressure Pipe,* Second Edition, 1995, #30009PA

M11, *Steel Pipe—A Guide for Design and Installation,* Fourth Edition, 1989, #30011PA

M12, *Simplified Procedures for Water Examination,* Fifth Edition, 2003, #30012PA

M14, *Recommended Practice for Backflow Prevention and Cross-Connection Control,* Second Edition, 1990, #30014PA

M17, *Installation, Field Testing, and Maintenance of Fire Hydrants,* Third Edition, 1989, #30017PA

M19, *Emergency Planning for Water Utilities,* Fourth Edition, 2001, #30019PA

M21, *Groundwater,* Third Edition, 2003, #30021PA

M22, *Sizing Water Service Lines and Meters,* First Edition, 1975, #30022PA

M23, *PVC Pipe—Design and Installation,* Second Edition, 2002, #30023PA

M24, *Dual Water Systems,* Second Edition, 1994, #30024PA

M25, *Flexible-Membrane Covers and Linings for Potable-Water Reservoirs,* Third Edition, 2000, #30025PA

M27, *External Corrosion—Introduction to Chemistry and Control,* First Edition, 1987, #30027PA

M28, *Rehabilitation of Water Mains,* Second Edition, 2001, #30028PA

M29, *Water Utility Capital Financing,* Second Edition, 1998, #30029PA

M30, *Precoat Filtration,* Second Edition, 1995, #30030PA

M31, *Distribution System Requirements for Fire Protection,* Third Edition, 1998, #30031PA

M32, *Distribution Network Analysis for Water Utilities,* First Edition, 1989, #30032PA

M33, *Flowmeters in Water Supply,* First Edition, 1989, #30033PA

M36, *Water Audits and Leak Detection,* Second Edition, 1999, #30036PA

M37, *Operational Control of Coagulation and Filtration Processes,* Second Edition, 2000, #30037PA

M38, *Electrodialysis and Electrodialysis Reversal,* First Edition, 1995, #30038PA

M41, *Ductile-Iron Pipe and Fittings,* Second Edition, 2003, #30041PA

M42, *Steel Water-Storage Tanks,* First Edition, 1998, #30042PA

M44, *Distribution Valves: Selection, Installation, Field Testing, and Maintenance,* First Edition, 1996, #30044PA

M45, *Fiberglass Pipe Design,* First Edition, 1996, #30045PA

M46, *Reverse Osmosis and Nanofiltration,* First Edition, 1999, #30046PA

M47, *Construction Contract Administration,* First Edition, 1996, #30047PA

M48, *Waterborne Pathogens,* First Edition, 1999, #30048PA

M49, *Butterfly Valves: Torque, Head Loss, and Cavitation Analysis,* First Edition, 2001, #30049PA

M50, *Water Resources Planning,* First Edition, 2001, #30050PA

M51, *Air-Release, Air/Vacuum and Combination Air Valves,* First Edition, 2001, #30051PA

To order any of these manuals or other AWWA publications, call the Bookstore toll-free at 1-(800)-926-7337.

This page intentionally blank.